Real Estate and Urban Development in South America

Real Estate and Urban Development in South America uncovers how investors are navigating South American real estate markets in commercial, residential and infrastructure development. A preferred location for real estate development during the colonial era, in recent decades South America has been seen as high-risk for global real estate investors. This book explores the strengths and weaknesses of real estate markets in the region, concluding that with careful implementation of the correct development strategies, the region can once again take its place at the centre stage of global real estate investment.

Comparing the economics and market maturity of South American countries in turn, the authors draw out the particular contexts in which investors and developers operate in mature and emerging markets. Bridging the gap between theory and practice, analysis of local development policies, legislation, valuation methods and taxation is supplemented with case studies from key players in the region's major cities. The first full overview of real estate markets in South America, this book will be an essential guide for investors, policy makers, academics and students with an interest in this this rapidly evolving region.

Claudia Murray is the Latin American convenor for the Academy of Urbanism and research fellow at the University of Reading, UK, where she is a member of the Group for the Americas. She is also fellow of the RSA and the Walker Institute for Climate Change.

Eliane Monetti is a professor at the University of São Paulo's Real Estate Research Group, treasurer for the Latin American Real Estate Society and a member of the World Economic Forum's Industry Agenda Council on Real Estate and Urbanisation.

Camilla Ween is Director of Goldstein Ween Architects, a Harvard Loeb fellow registered at the ARB, member of the RIBA, fellow of the RSA, member of the CIHT and academician with the Academy of Urbanism.

International Real Estate Markets

The Chinese Real Estate Market
Development, Regulation and Investment
Junjian Cao

Real Estate in Italy
Markets, Investment Vehicles and Performance
Guido Abate and Guiditta Losa

Real Estate and Urban Development in South America
Understanding Local Regulations and Investment Methods
in a Highly Urbanised Continent
Claudia Murray, Eliane Monetti and Camilla Ween

Real Estate and Urban Development in South America

Understanding Local Regulations
and Investment Methods in a Highly
Urbanised Continent

**Claudia Murray, Eliane Monetti
and Camilla Ween**

Routledge
Taylor & Francis Group

LONDON AND NEW YORK

First published 2018
by Routledge
2 Park Square, Milton Park, Abingdon, Oxon OX14 4RN

and by Routledge
711 Third Avenue, New York, NY 10017

Routledge is an imprint of the Taylor & Francis Group, an informa business

British Library Cataloguing-in-Publication Data
A catalogue record for this book is available from the British Library

Library of Congress Cataloging-in-Publication Data
Names: Murray, Claudia, author. | Monetti, Eliane, author. | Ween,
    Camilla, author.
Title: Real estate and urban development in South America :
    understanding local regulations and investment methods in a highly
    urbanised continent / Claudia Murray, Eliane Monetti and Camilla
    Ween.
Description: Abingdon, Oxon ; New York, NY : Routledge, 2018. |
    Includes bibliographical references and index.
Identifiers: LCCN 2017033209 | ISBN 9781138185500 (hardback :
    alk. paper) | ISBN 9781138185517 (pbk. : alk. paper) | ISBN
    9781315644455 (ebook : alk. paper)
Subjects: LCSH: Real estate development—South America. | Real estate
    investment—South America. | City planning—South America. |
    Urbanization—South America. | Rural development—South America.
Classification: LCC HD466 .M87 2018 | DDC 333.33098—dc23
LC record available at https://lccn.loc.gov/2017033209

ISBN: 978-1-138-18550-0 (hbk)
ISBN: 978-1-138-18551-7 (pbk)
ISBN: 978-1-315-64445-5 (ebk)

Typeset in Goudy
by Apex CoVantage, LLC

Contents

Figures

Graphics

Tables

Boxes

Case study

Contributors

Claudio Tavares de Alencar is a member of the University of São Paulo Polytechnic School Real Estate Research Group. He gained his PhD in 1998. His research in real estate deals with the following themes: planning, management, business, marketing, economics, funding and sustainability. More precisely, his work focused on the development of systems that allow the improvement of the whole decision process in the real estate sector. In his studies, the upgrade of the decision making process is supposed to be conducted by an advancement in the knowledge related to planning techniques involved in all segments of the management of the sector. This takes into account the main features of real estate and the particularities of its economic insertion and impacts on sustainability, as well as considering the dimension and duration of real estate projects, the drives behind the planning methods and routines, emphasising the risk analysis of the projects and companies. He is also a former President of the Latin American Real Estate Society (LARES).

Edmund Amann is Professor of Brazilian Studies at Leiden University and Visiting Professorial Lecturer at the School for Advanced International Studies, Johns Hopkins University. Previously he was Reader in Development Economics at the University of Manchester and a Research Fellow at the University of Oxford. His research centres on regulation, innovation and foreign direct investment in a developing country context. Much of his work focuses on the experiences of Latin America, especially Brazil. He has published in a wide range of development and economics journals, including *World Development* and *Oxford Development Studies*. In addition, he has acted as a consultant for the Inter-American Development Bank.

Jim Costello has worked in the CRE space on issues of urban economics since 1990, including a 20-year stint with CBRE's Torto Wheaton Research team. He expanded the reach of Torto Wheaton Research developing forecasts of global market fundamentals. He also developed frameworks to interpret the forecasts in answer to investor questions on asset values and relative investment opportunities. He provided advice to the Treasury Department in the aftermath of the global financial crisis and helped educate these professionals on commercial real estate performance. He is a member of the Commercial Board of Governors of the Mortgage Bankers Association and is working there to help leading lenders understand how economic forces impact loan performance. He is expanding the capabilities of the Real Capital Analytics team on issues of real estate market dynamics. He has a master's degree in economics and is a member of the Counselors of Real Estate.

Nicolas Estupiñan served as transport vice-minister in Colombia during 2012–2014. He holds a degree in civil engineering from Universidad de los Andes in Bogotá, and a master's in City Planning and Transportation from the University of North Carolina at Chapel Hill. During his 15 years of experience, he has worked for the public sector in Colombia and for development banks enhancing the urban transport agenda, fostering sustainable development and poverty reduction through applied research, program and project design and implementation. He has published several works on urban transport affordability and urban form affecting travel patterns. He serves for three committees at the Transport Research Board, and is a member of the Michelin Stakeholder Committee. Currently, he works for CAF – Development Bank of Latin America as senior transport specialist, based in Buenos Aires, coordinating the urban transport group.

German C. Lleras has worked as an academic, public official and consultant in areas related to cities, infrastructure and transport. During the last 20 years, he has provided strategic advice to city administrations, communities and the private sector, particularly in Latin America. His work has helped in designing policies and projects like Transmilenio in Bogota and has played an important role in the financing of several transport infrastructure projects. He currently is the Steer Davies Gleave regional director for Latin America and a lecturer at the Civil and Environmental Engineering at Universidad de los Andes in Bogota.

Sebastian Piliponsky is a civil engineer from the National University of Tucumán. He has several master's degrees related to real estate development including a master's in Management and Organisation of Construction Companies from the Universidad Politécnica de Catalunya and a master's in Project Management from the Universidad Ramón Llull, also in Catalunya. He was Director of ISAURA, a developer in Barcelona from 2001 to 2011 and today he is a partner of LINK investment in Argentina (since 2011). He is currently developing 13 projects of approximately 65,000 m^2 in Tucumán, Argentina. Academically he is a professor at ENRE (www.enrealestate.org) and co-author of the book *The Art of Financing Real Estate Projects* and editor of the book *Tucumán under Construction*, both published in Argentina.

Lake Sagaris is an expert on cycle-inclusive urban planning, civil society development and participatory urban-regional governance, planning and practice. An award-winning writer and editor, she has worked in Chile since 1980 as a freelance journalist. She holds a master of science degree (University of

Toronto, 2006) and a PhD in Urban Planning and Community Development (University of Toronto, 2012). Her current work uses participatory action research methods and community-government partnerships in Santiago and Temuco (Chile) to apply an intermodal approach based on "ecologies of modes and actors" to transition toward more sustainable transport. She focuses on resilience, social justice and inclusion, particularly gender, safety and security issues. These experiences have led to awards in Chile and abroad, participation in a UN Expert Group meeting on sustainable transport, and presentations in Latin America, Europe, Canada, Taiwan, the US, and India.

Introduction

South America's institutions and culture affecting real estate investment: can path dependency be broken?

South America presents a history of economic and socio-political upheavals that have stigmatised the region as high-risk for global real estate investors. Even to this day and after nearly 30 years of reasonably stable democratic governments, key barriers for international RE investment still remain in place in some countries. Venezuela, for example, has currency restrictions while the more liberal Chile has capital controls.

The region was not always seen as hostile. In fact, real estate development and investment is what shaped many cities during the times of the colonies (Murray, 2008), and even in post-colonial periods and up to the beginning of the First World War, South American countries attracted international investment for infrastructure that helped to develop sophisticated investment vehicles particularly from Great Britain (Nurkse, 1954). So, historically, the region has a longer record for being an exciting location for international investors than for being a high-risk market. Notwithstanding, the characterisation as a volatile region is the prevailing one, while countries of the Pacific Rim, which a century ago were bypassed by investors, are today favoured by them (Fukuyama, 2008). Paradoxically, the reasons behind this categorisation of a risky region in South America have more to do with its long-term history (and therefore that of a good place for real estate investment) than the bad turnings and political decisions taken during the last century.

Path dependence is considered by many academics an important factor in the development of state policies (Malpass, 2011; Bengtsson and Ruonavaara, 2010). Widely speaking, the theory maintains that the existing structures of institutions and practices in any country frame the unique nature of its government and influence its direction of travel and the particular shape that the change takes. For Latin America as a whole, some authors argue that latifundia practices imposed by the colonial rulers and the subsequent failure from different governments to distribute land in an equitable way, has put countries under pressure from a variety of factors (trade unions, political activists, intellectuals, indigenous communities) that are now intensifying their claims to access land (Fukuyama, 2008). This has left the region with a fundamental deficiency in many land titles that some claim needs to be corrected in order to effectively create a sustainable property rights environment that can help not only development, but free the barriers for the collateralisation of assets for entrepreneurial investment (de Soto, 2001; for a criticism of this view see Gilbert, 2002b).

It can be argued that title deficiency is an issue applicable to urban and peri-urban informal developments only, but there are cases of land grabbing where small rural farmers have been affected. Indeed there are examples in Bolivia, Peru and most notably Colombia, where many have been forced out of their lands by the drug wars (Borras et al., 2012; Murray, 2015). This deficiency creates serious difficulties particularly when countries attempt to develop large infrastructure projects across areas of conflict. Equally problematic is the development of an infrastructure around large urban agglomerations, where informal developments also create confrontations. Many times these conflicts can reach worldwide attention, particularly when natural resources are at risk, as it was the case of Brazil's Belo Monte dam in the Amazon basin.[1] In more urbanised contexts, the conflict hardly reaches the international media, but can be the cause of road closures and pickets by so-called "illegal" residents. As a consequence, most governments and politicians are usually not willing to enter into conflicts with these groups, particularly when the said groups have the support of powerful trade unions.

This fundamental land conflict is coupled with an inherited patrimonial mentality that equates land to power. Indeed this philosophy materialised clearly in the Tordesillas Treaty signed in 1494, which divided all new discoveries of land outside Europe between the Portuguese and Spanish Empires. This patrimonial mentality drove the colonisation of the territories of Brazil and rest of Latin America thereafter, intensifying in post-colonial times, when the region followed a global tendency towards individual ownership of land. It was also this patrimonial mentality that was the underlying force behind the creation of cities and the establishment of an urban network in what is today the most urbanised continent on earth. However, the patrimonial mentality was always in conflict with traditional forms of communal landownership followed by most South American indigenous groups (Griffiths, 2004). Current issues around food, energy and water security are adding extra pressure to the land conflict in Latin America as a whole, as the region revisits its past as a global exporter of commodities (Borras et al., 2012). But the problem is not only about rural land, there is also an internal migration triggered by this land grabbing process that directly affects urban areas.

Indeed for the past five decades, the rural population has been steadily moving towards bigger urban centres and the region now has one of the highest urban population rates in the world; for example, Uruguay's population is 95% urban. A highly urbanised region might sound beneficial for real estate but this particular migration has tended to settle in informal developments, adding to the national housing deficits and infrastructure crisis (Dufour and Piperata, 2004; Murray and Clapham, 2015). There are other push factors for this influx of migrants to the cities, such as the exploitation of natural resources, that can cause environmental damage, with water pollution being the most notable as the case of Peru demonstrates (Bebbington and Williams, 2008). Water shortage has a direct effect on the livelihoods of small farmers and many are forced to move and find work in nearby towns. Natural disasters are another cause for migration as well as local armed conflict. The current Colombian housing programme of "100,000 free homes" (Cien Mil Viviendas Gratis), for example, was designed to deal precisely with environmental and war migrants from the rural areas (Murray, 2015). Migration is inevitable, and sometimes good for real estate as housing mobility generates market demand. However, the promise of property ownership (as opposed to renting) by local governments is harming the residential market. Indeed this becomes evident when much of what is being offered to potential buyers is of such low standards that some claim construction companies delivering these units would be out of business if buyers truly had a choice outside the subsidised government systems (Formoso et al., 2011).

In addition to the land conflict and patrimonial mentality, some argue that the region has a weak judicial system and ineffective tax collection mechanisms that tend to benefit the elite (Chomsky, 2010). These factors exacerbate social volatility and create the ideal environment for ideological battles of extreme left and right movements. Battles are sometimes fought by the locals or by a constant stream of international intellectuals whose partisan theories ignite negative politics, keeping alive the romantic idea of a land of revolutionaries, which can have a harmful effect on foreign investors' perception of the region. One such international group is the anti-globalisation movement embodied by the Social Forum, which gathers civil society groups from the global North and South in annual meetings. The Forum has a considerable force in South America, with Porto Alegre in Brazil being the location for the first-ever gathering of the group.

As demonstrated by some (Gwynne and Cristobal, 2016), globalisation is at the centre of a Latin American debate that sees some countries benefiting from the process while some others are falling behind. In other words the expected convergence is not occurring as predicted by the supporters of global integration (Held et al., 1999), and some South American countries, such as Bolivia, Ecuador and Venezuela, are amongst the most vociferous detractors of global integration. At the country level, this battle of extremes and negative politics damage the democratic process of dialectics and conciliation at all levels of government, from ministries to local municipalities. This leads to unhappy outcomes such as the almost dictatorial state of Venezuela, which has been dominated by the United Socialist Party of Venezuela (PSUV for its Spanish Acronym) since the ascent of former president Hugo Chavez (†2013) in 1999. His successor, Nicolas Maduro, narrowly won the elections, allegedly using State funds to finance his political campaign (Lopez Maya, 2014). Since then, the polarisation of politics has worsened in the country, with claims that the PSUV is persecuting all opposition parties under the excuse of a potential a coup d'état instigated by them. At the time of writing, the country is suffering a wave of anti-government demonstrations that are escalating into violence, fuelled by high inflation, a food crisis and medicine shortages.

Other countries are also suffering from this polarisation of politics, albeit in less extreme fashion than Venezuela. The new president of Argentina, Mauricio Macri, managed to win the elections in 2015 with his party Cambiemos, a coalition of centrist parties that concluded 15 years of Peronist ruling. This signified a complete U-turn in politics, moving from a heavily centralised populist and inward-looking government, to a pro-business government that has embarked in widespread reforms, including the reduction of popular subsidies (such as those in the energy sector), which considerably shortened households' disposable income. In a country that is facing the highest inflation in South America outside Venezuela, the reforms have been received by a wave of protests from opposition groups as well as state workers and trade unions.

At the macro level, left and right ideological battles can change the degree of openness and liberalisation of markets to foreign investors from one presidential election to the next, depending on the views of the elected parties, as the extreme left usually imposes high restrictions to foreign investment. Considering that foreign direct investment in real estate is mostly long term, international investors usually shy away not only from deficient titles but also from judicial systems and rule of law that are at the mercy of extreme politics and can change the market conditions overnight.

If not threatened by political extremes and abuse of power, South American countries can suffer from corruption scandals. The impeachment of Dilma Rousseff in Brazil (2015) not only ousted the incumbent president but also kicked off an investigation into the state-owned oil company Petrobras. A current investigation involves 83 politicians over allegations of over-priced contracts at the oil firm.[2] Furthermore, the Brazilian construction company Odebrecht is also at the centre of infrastructure contract bribes that has ensnared political figures in other South American countries, the most notable example being Peru, involving the former President Alejandro Toledo, who is accused of receiving over USD 20 million for the granting of a contract to build a transoceanic highway between Brazil and the Peruvian coast.[3]

Title deficiency, partisan politics and corruption scandals can add risk to long-term real estate investment. For the future and with these antecedents, the course of path dependency theories in South America presents a bleak picture. Notwithstanding, there are criticisms to these theories that must be brought to light. It is argued (Kay, 2005) that there are some general forces, such as globalisation, that are moving systems in similar directions (convergence) and, therefore, national policies are constantly being changed. One such fundamental change happened during the 1990s when most South American countries adopted important economic recommendations known as the Washington Consensus. These new policies were based on a neo-liberal ideology proposed by the World Bank (World Bank, 1993; Gilbert, 2002a; Stiglitz, 2003). Broadly speaking, this implied the reduction of the state, decentralisation, privatisation, opening of markets and deregulation. The adoption of these recommendations has had a direct impact on real estate investment. Unfortunately, lack of transparency in the privatisation process and the lack of a mature institutional framework capable of dealing with the sophistication that international private investment demands have together been blamed for the economic crises that affected the region after market liberalisation, starting with Mexico's Tequila crisis in 1994, Brazil's hyper-inflation crisis of 1994 and ending with Argentina's default crisis in 2001 (Stiglitz, 2007). So it can be said that the South America of the 1990s was still haunted by its institutional past. But the case of Chile is enough to exemplify that path dependency can be dramatically changed. The difference here was that Chile adopted the Washington Consensus at a much earlier stage (around the 1970s) and therefore had a longer term to adjust to economic reforms. Chile is today the only South American country that is an OECD member and expected to become the first developed country in the region (Larroulet, 2013). Still most countries, including Chile, are highly dependant on international finances and, in particular, US investment in the region (ECLAC, 2004). As it has been pointed out, this dependency on foreign markets not only leaves the local currencies to the mercy of capital flight and fluctuations (as when FDI rises, local currencies also rises, which in turn damages exports), but also leaves the region with very little room to manoeuvre in terms of setting up policies that can be upsetting to foreign investors (Gwynne and Cristobal, 2016).

The good economic conditions enjoyed by most countries during the first decade of the new millennium have helped to strengthen the real estate sector and its regulatory framework. The most notable example is Brazil, which has introduced a series of urban tools in the city of São Paulo that have improved land value capture mechanisms for the development of infrastructure and social housing. Furthermore, in 2013 the Brazilian government exempted foreign investors from paying a 6% financial transaction tax on purchases of real estate investment trusts traded on the country's stock exchange (the Brazilian FII – Fundo de Investimento Imobiliário). This was a first step to increasing monetary flows from overseas, especially from the US, which has led to a surge of initial public offerings (IPOs) of Brazilian construction firms. In addition Brazil has also received a lot of attention over the past five to ten years with growing demand for hotels and offices due to global events being hosted in Brazil, including the World Cup in 2014 and the summer Olympics in 2016. Despite the increase of dollar inflows, Brazil's government-backed social housing schemes such as Minha Casa, Minha Vida struggle to be successful as the lower construction costs do not make up for the lower margins construction companies would receive for these schemes. Thus, the only way to achieve a reasonable margin is to compromise heavily on quality standards. Residential market has also developed, raising the ratio of real estate financing to the GDP from 2% to 9% in less than a decade term.

The current century presents a new opportunity for the rest of the region to change the course of path dependence and improve on recent achievements. Real estate and infrastructure development is at the centre stage of this opportunity. As the Economic Commission for Latin American and the Caribbean (ECLAC) has stated, the key is to implement sustainable growth. Fundamental to sustainable development is the provision of adequate and affordable housing for all sectors of the population; delivery of key infrastructure that will sustain a reasonable quality of life and support access to jobs and opportunity. These actions must all properly embrace climate change mitigation strategies and environmental protection. In the RE context, this means to preserve natural resources and share prosperity with all sectors of the population while simultaneously creating the ideal conditions for foreign and national investment. Most countries in South America have pledged to fundamental changes through their Millennium Development Goals and are making considerable progress.

In these new Millennium conditions, this book intends to uncover how investors are managing to navigate the real estate markets in commercial, residential and infrastructure development. By bringing case studies from a variety of countries, written by those at the front line of the development, the authors will provide a current picture of real estate markets in South America. The methodology aims to uncover strengths and weaknesses of RE markets and urban development in the region, with the final objective of providing a strong policy framework that, if adopted by governments, can bring confidence to the market and place the region again at the centre stage of global real estate investment.

The book is arranged as follows: Chapter 1 presents an overview of South American countries, including a ranking of the strongest economies in terms of their RE markets potential. Chapter 2 covers commercial RE, focusing on the office and retail sector. Chapter 3 presents the residential market, including social housing policies and delivery. Chapter 4 brings an analysis of the different investment vehicles available as well as financing mechanisms for the development of commercial and residential properties. Chapter 5 presents the infrastructure sector, evaluating how state investment drives countries' productivity and GDP. Chapter 6 shows how politics, culture and economic cycles affect inward investment from the wider Latin American economies and other world regions. The final Chapter (7) concludes with an analysis of real estate development in the context of the Sustainable Development Goals and the Agenda 2020 that were signed by most countries under study in this volume.

Chapters 4, 5 and 6 include contributions from professionals who are at the forefront of real estate development in South America. Chapter 4 has one contributor from Brazil (Expert box 4.1) who explains the functioning of the Brazilian REIT, and a contribution from a real estate developer in Argentina (Expert box 4.2) who provides his view on current financing systems for housing in a country characterised by high levels of inflation. Chapter 5 has four contributors who assess the state of infrastructure in Argentina, Brazil, Colombia and Chile. Chapter 6 has a contribution from Real Capital Analytics, a global firm that specialises in data collection and analysis for real estate transactions at the global level.[4] The inclusion of these experiences from those in practice directly informs each chapters' conclusions and final comments for this volume.

Notes

1 www.theguardian.com/world/2016/jan/15/brazil-belos-monte-dam-delay-court-indigenous-people
2 www.ft.com/content/6e8b0e28-f728-11e5-803c-d27c7117d132?mhq5j=e1
3 www.ft.com/content/cd885b26-eb1d-11e6-930f-061b01e23655?mhq5j=e1
4 www.rcanalytics.com/company/about-rca/

Bibliography

Bebbington, A. and Williams, M. (2008) Water and Mining Conflict in Peru. *Mountain Research and Development*, 28, 3–4, pp. 190–195.

Bengtsson, B. and Ruonavaara, H. (2010) Path Dependence in Housing: Introduction to the Special Issue. *Housing, Theory and Society*, 27, 3, pp. 193–203.

Borras, S., Franco, J., Gomez, S., Kay, C., and Spoor, M. (2012) Land Grabbing in Latin America and the Caribbean. *Journal of Peasants Studies*, 39, 3–4, pp. 854–872.

Chomsky, N. (2010) *Hopes and Prospects*. London: Hamish Hamilton.

de Soto, H. (2001) *The Mystery of Capital: Why Capitalism Triumphs in the West and Fails Everywhere Else*. London: Black Swan.

Dufour, D. and Piperata, B. (2004) Rural-to-Urban Migration in Latin America: An Update and Thoughts of the Model. *American Journal of Human Biology*, 16, 4, pp. 395–404.

ECLAC. (2004) *Foreign Investment in Latin America and the Caribbean 2003*. Available at www.cepal.org/en/publications/1127-foreign-investment-latin-america-and-caribbean-2003, accessed on 8th July 2016.

Formoso, C., Leite, F., and Miron, L. (2011) Client Requirements Management in Social Housing: A Case Study on the Residential Leasing Programme in Brazil. *Journal of Construction in Developing Countries*, 16, 2, pp. 47–67.

Fukuyama, F. (2008) *Falling Behind: Explaining the Development Gap Between Latin America and the United States*. Oxford: Oxford University Press.

Gilbert, A. (2002a) Power, Ideology and the Washington Consensus: The Development and Spread of the Chilean Housing Policy. *Housing Studies*, 17, 2, pp. 305–324.

Gilbert, A. (2002b) On the Mystery of Capital and the Myths of Hernando de Soto: What Difference Does Legal Title Make? *International Development and Planning Review*, 24, 1, pp. 1–19.

Griffiths, T. (2004) Indigenous Peoples, Land Tenure and Land Policy in Latin America. *Land Reform*, 1, pp. 46–63.

Gwynne, R. and Cristobal, K. (2016) *Latin America Transformed: Globalisation and the Modernity*. London: Routledge.

Held, D., McGrew, A., Goldblatt, D., and Perraton, J. (1999) *Global Transformations: Politics, Economics and Culture*. Cambridge: Polity Press.

Kay, A. (2005) A Critique of the Use of Path Dependence in Policy Studies. *Public Administration*, 83, 3, pp. 553–571.

Larroulet, C. (2013) Chile's Path to Development: Key Reforms to Become the First Developed Country in Latin America. *The Heritage Foundation*. Available at www.heritage.org/research/reports/2013/10/chiles-path-to-development-key-reforms-to-become-the-first-developedcountry-in-latin-america, accessed on 16th April 2015.

Lopez Maya, M. (2014) Venezuela: The Political Crisis of Post-Chavismo. *Social Justice*, 40, 4, pp. 68–87.

Malpass, P. (2011) Path Dependency and the Measurement of Change in Housing Policy. *Housing Theory and Society*, 28, 4, pp. 305–319.

Murray, C. (2008) The Regulations of Buenos Aires' Private Architecture During the Late Eighteenth Century. *Architectural History*, 51, pp. 137–160.

Murray, C. (2015) Real Estate and Social Inequality in Latin America: Approaches in Argentina, Brazil, Chile and Colombia. In Abdulai, R., Obeng-Odoom, F., Ochieng, E., and Maliene, V. (eds.), *Real Estate, Construction and Economic Development in Emerging Market Economies*. London: Routledge, pp. 256–276.

Murray, C. and Clapham, D. (2015) Housing Policies in Latin America: Overview of the Four Largest Economies. *International Journal of Housing Research*, 15, 3, pp. 347–364.

Nurkse, R. (1954) International Investment Today in the Light of the Nineteenth Century Experience. *The Economic Journal*, 64, 256, pp. 744–758.

Stiglitz, J. (2003) *Globalisation and Its Discontents*. London: Penguin Books.

Stiglitz, J. (2007) *Making Globalisation Work*. London: Penguin Books.

World Bank. (1993) *Housing: Enabling Markets to Work*. Washington, DC: World Bank.

1 Comparative overview of key real estate drivers

1.0 Introduction

Research and thought process for this book were developed between 2015 and the first quarter of 2017. At the time of writing, the global economy remains in transition since the Global Financial Crisis (GFC) of 2008, and political changes are happening in key countries that can have a direct influence on South America. Indeed, there are certain aspects of the global economy that have always influenced the economies of the countries under study here, namely the state of the US economy and the commodities market. Additionally, the increased integration of China to the global scene is a relatively new added factor as the slowdown in manufacturing in China means a reduction in the demand for the region's raw materials. Furthermore, many South American governments are building large debts to the Asian giant (Table 1.1), which adds to the region's vulnerability to the latter's current slower but more sustainable pace of growth.

High levels of debt do not bode well with this new reality of slow growth, particularly since many South American countries enjoyed the windfalls of the regional boom of 2009–2011. The good years were spent in building and implementing large social programmes funded by the states (Murray and Clapham, 2015), but little was set aside to create a buffer for the bad years. The new wave of more liberal governments are therefore trapped in unsustainable programmes which are very unpopular to dismantle and very difficult to approve through the respective congresses. As this chapter will demonstrate, the regional economic hierarchy of countries is also undergoing a reshuffle in alignment with political changes briefly described earlier in the introduction, and the winners are those whose policymakers are most quick to respond to the new world's reality of slow growth and low commodity demand.

In South America as in the rest of the world, the state of the economy and key market fundamentals have a significant impact on real estate. However, the peaks and troughs of an economic cycle can be balanced out over the lifetime of a long-term investment such as property, while other short-term investments such as equities can carry greater volatility in South America than in more stable developed economies.

Looking at key social and economic indicators, this chapter provides the basis of the present volume and defines the economic and real estate market framework that will be used to explain the particularities of South America. The framework

Table 1.1 South America's debt to China

Country	Amount
Venezuela	USD 62.2 billion
Brazil	USD 36.8 billion
Argentina	USD 15.3 billion
Bolivia	USD 3.5 billion
Peru	USD 50 million

Source: The Dialogue database, www.thedialogue.org/map_list

Data: Total debt in USD since 2005

presented will enable the reader to understand real estate market analysis as a whole and its application to each country under study. In order to build this analytical framework, internal and external indicators affecting the property market have been used and explained. Thus, the indicators will include a combination of macro and microeconomic factors as well as fundamental real estate market variables. Using the latest data from reliable open source databanks (World Bank, United Nations, IMF, etc.), Section 1.1 starts by presenting the comparative statistics in relation to demographics, population growth and migration for each country, placing them in context with two advanced economies: UK and the US.

The next section (1.2) presents the economic drivers of the region including: GDP, unemployment data, and urban growth. This will help to complete the wider picture of the level of development and urbanisation of countries. After a selection of the most important economies, the following section (1.3) brings crucial market data such as interest rates, inflation, wages, consumer spending as well as general supply and demand statistics. Section 1.4 looks at the planning system and legal framework while the final section (1.5) brings the conclusion and evaluates and divides the countries according to the importance of their real estate markets in a global context (mature South American markets) and according to their possibilities (emerging South American markets).

1.1 Key real estate drivers

1.1.1 Demographics

Demographics represent an important piece of data that provides information on the state of the population, its growth and composition. Analysts can infer many aspects of the state of real estate markets through such information; for example, they can tell if the demand for smaller houses is higher due to a rise in new families; or measure the impact that migration can have in property prices as the number of homes needed increases.

The 12 sovereign states that comprise South America have a total population of just over 418 million, which represents nearly two-thirds of the 637 million of Latin Americans that inhabit the region (Population Reference Bureau, 2016). Brazil alone contributes to nearly a third of the grand total (Table 1.2),

Table 1.2 Total population (in millions)

Argentina	43.6
Bolivia	11
Brazil	206.1
Chile	18.2
Colombia	48.8
Ecuador	16.5
Guyana	0.8
Paraguay	7
Peru	31.5
Suriname	0.5
Uruguay	3.5
Venezuela	31
Total	418.5
UK	65.6
US	323.9

Source: Population Reference Bureau – 2016 world population data sheet www.prb.org/pdf16/prb-wpds2016-web-2016.pdf

Data: Population (total) for (mid) 2016

while the rest of the countries can be grouped as: Colombia and Argentina in the 40 million mark, Peru and Venezuela in the 30 million, and the remaining seven countries are under 20 million. The population of the UK and the US are also provided for reference, with 65.6 and 323.9 million inhabitants, respectively. Brazil, as well as the US, are amongst the five most populated countries in the world, while Colombia and Argentina have less than the UK.

Demographics can be described through many factors (gender, age, ethnicity), but for real estate markets the most relevant is growth rate. This is mainly because this helps to forecast the demand for built space (residential/commercial) needed by countries.

1.1.2 Population growth

During the past ten years, South America has had an average population growth rate of 11.6% (see Table 1.3). Countries such as Bolivia and Ecuador have enjoyed a growth rate of over 20%, while Uruguay is much lower with 6.1% and Suriname even less, showing no population variance since 2006.

In terms of forecasting, the regional average shows a slight increase of 1% (Table 1.4), indicating that the regional growth rate remains roughly the same as the past ten years shown in Table 1.3. The exception is Colombia, which jumps from a recorded growth rate in the past ten years of 4.3% to a forecasted 9.2% in the next 15 years. This constitutes a rate hike of 0.4% to a 0.6%, which can be explained by the end of the armed conflict and the country's confidence in a successful peace process with remaining rebel groups (see Introduction). The only other country with a jump in population growth rate is Suriname, which is predicted to go from no growth in the past ten years to an average 1.3% a year until

Table 1.3 Population growth (%) for the last ten years up to 2016

	2006 (in millions)	2016 (in millions)	Growth (%) [authors' own calculations]
Argentina	39	43.6	11.8%
Bolivia	9.1	11	20.9%
Brazil	186.8	206.1	10.3%
Chile	16.4	18.2	11.0%
Colombia	46.8	48.8	4.3%
Ecuador	13.3	16.5	24.1%
Guyana	0.7	0.8	14.3%
Paraguay	6.3	7	11.1%
Peru	28.4	31.5	10.9%
Suriname	0.5	0.5	0.0%
Uruguay	3.3	3.5	6.1%
Venezuela	27	31	14.8%
Total	377.6	418.5	10.8%
Average			11.6%
UK	60.5	65.6	8.4%
US	299.1	323.9	8.3%

Source: Population Reference Bureau – 2006 world population data sheet www.prb.org/pdf06/06 WorldDataSheet.pdf. Growth rate calculated by authors

Data: Population growth (%) for the last ten years up to 2016

Table 1.4 Population growth (%) forecast up to 2030

	2016 (in millions)	2030 (in millions)	Growth (%) [authors' own calculations]
Argentina	43.6	49.1	12.6%
Bolivia	11	13.3	20.9%
Brazil	206.1	223.1	8.2%
Chile	18.2	20.3	11.5%
Colombia	48.8	53.3	9.2%
Ecuador	16.5	19.7	19.4%
Guyana	0.8	0.8	0.0%
Paraguay	7	8.2	17.1%
Peru	31.5	35.9	14.0%
Suriname	0.5	0.6	20.0%
Uruguay	3.5	3.6	2.9%
Venezuela	31	36.1	16.5%
Total	418.5	464	10.9%
Average			12.6%
UK	65.6	71	8.2%
US	323.9	359.4	11.0%

Source: Population Reference Bureau – 2016 world population data sheet www.prb.org/pdf16/prb-wpds2016-web-2016.pdf. Growth rate calculated by authors.

Data: Population growth (%) forecast up to 2030

2030, giving a total increase of 20% above the population recorded in 2016. However, other sources are more conservative about Suriname's population expansion, forecasting an annual rate of 0.7% (see, for example, World Bank's World Development Indicators and IMF's World Economic Outlook Database, 2016).

In summary, all South American countries are showing a slower rate of population growth for the next 15 years compared to the last ten up to 2016, with the exception of Colombia and Suriname with a conservative estimate of 0.6% and 0.7% growth rate per year, respectively.

1.1.3 Migration

Free movement of people from one country to another affects the population growth rate. A particular factor influencing demographical changes in South America is the creation of the common market known as Mercosur (1991) and the Andean Community of Nations – CAN (1969). Full members of Mercosur include: Argentina, Brazil, Paraguay, Venezuela and Uruguay, while associate members are Bolivia, Chile, Peru, Colombia, Ecuador and Suriname.[1] On the other hand, CAN's main members are Bolivia, Colombia, Ecuador and Peru. In 2005, CAN extended its membership to all other main Mercosur members. In addition to these two groups, a new block known as Union of South American Nations was established in 2008. This new group joins full members of Mercosur, full members of CAN as well as three new members: Chile, Guyana and Suriname. Irrespective of the trade block to which they belong, all 12 South American countries debate migration rules and agreements at the annual meeting of the South American Conference on Migration (CSM), created in 2000.[2]

As in the European Union, the aim of all South American alliances is building collaboration and to promote free movement of goods, people and currency. But unlike the EU, the outcomes of the CSM are non-binding and the group does not have a supranational body (like the EU Court of Justice) to implement migration agreements. Nor has it an abolition of borders treaty such as Europe's Schengen area, although it recognises each other countries' national identification cards as travel documents.

Notwithstanding, the most effective agreement in terms of migration is the Mercosur Residence Agreement (2002). This accord allows migrants the right to reside and work for a period of two years in another signatory state, provided they hold a Mercosur country citizenship and have a clean criminal record. After two years of temporary residency migrants can, if they wish, remain in the host country and obtain a permanent residency. Guyana, Suriname and Venezuela have still to implement the Agreement, but the remaining nine countries have done so. As a result, between 2004 and 2013, there were nearly 2 million South Americans who obtained a temporary residence permit in another Mercosur country, with Argentina, Chile and Brazil seeing the largest increase in permits granted each year (IOM, 2014).

The Mercosur Residence Agreement can be seen as an attempt from all governments to control a growing trend of informal international migration that has seen record levels due to the displacement of many who are trapped in conflict, particularly Venezuela (see Introduction). The question here is how South American countries are coping with this surge in immigration.

1.1.4 Internal migration and the growth of informality

During the last decades, some countries like Argentina and Brazil have shown signs of slowing the rate of rural to urban migration, but others – Colombia and Peru, for example – are still catching up with the urbanisation trend of their neighbours. Table 1.5 shows latest available data from the UN Population Division for all South American countries, as well as the UK and US for comparison. Most countries have passed the 80% urban mark. For example, Argentina, Venezuela and Uruguay are amongst the most urbanised countries in the world (all above 90%), along with Belgium (98%), Iceland (94%) and Japan (93%), according to the same data source. It is worth pointing out that the level of urbanisation is not a sign of the country's wealth and state of development, and that the resources and technological advances needed to manage cities can certainly have an impact on a country's economy and quality of life of new city arrivals.

Table 1.5 Urban and rural population

	Urban	Rural	Total	Percentage Urban
South America	341,279	69,653	410,932	83.0
Argentina	38,293	3,510	41,803	91.6
Bolivia (Plurinational State of)	7,388	3,460	10,848	68.1
Brazil	172,604	29,429	202,034	85.4
Chile	15,881	1,892	17,773	89.4
Colombia	37,265	11,665	48,930	76.2
Ecuador	10,152	5,831	15,983	63.5
French Guiana	215	41	255	84.1
Guyana	229	575	804	28.5
Paraguay	4,110	2,807	6,918	59.4
Peru	24,088	6,681	30,769	78.3
Suriname	359	184	544	66.1
Uruguay	3,253	166	3,419	95.2
Venezuela (Bolivarian Republic of)	27,439	3,412	30,851	88.9
United Kingdom	52,280	11,209	63,489	82.3
United States of America	262,734	59,849	322,583	81.4

Source: United Nations, Population Divisions. World Urbanisation Prospects: The 2014 Revision at Mid-Year

Data: Population of urban and rural areas in thousands and percentage

Table 1.6 shows the percentage of urban population living in slums according to the UN Millennium Development Goals Indicators. With the exception of Argentina, Brazil and Chile, which as previously stated take the bulk of the regional international migration, it can be safely assumed that in the rest of the countries, most new city dwellers have materialised through internal migration. Whether this is rural-to-urban or urban-to-urban movement, there is clear evidence that certain countries are not managing city expansion as well as others. For example, Bolivia has 68.1% of urban population out of which 43.5% live in slums; equally, Guyana has 28.5% urban with 33.1% living below standards. Suriname, on the other hand, has a low level of urbanisation at 66.1% but its slum population seems to be on the increase, going from 3.9% in 2005 to 7.3% in 2014.

It remains to be researched whether it is rural to urban migration that drives the percentage of slums dwellers within these countries, but there is a clear correlation here between internal migration and low urban living conditions. As we will see later in this volume, this situation of populations living in informal conditions adds to the demand not only of new housing, but as well to that of the

Table 1.6 Urban population living in slums (%)

	1990	1995	2000	2005	2007	2009	2014
Argentina	30.5	31.7	32.9	26.2	23.5	20.8	16.7
Bolivia	62.2	58.2	54.3	50.4	48.8	47.3	43.5
Brazil	36.7	34.1	31.5	29	28	26.9	22.3
Chile				9			low proportion of slum dwellers*
Colombia	31.2	26.8	22.3	17.9	16.1	14.3	13.1
Ecuador				21.5			36
Guyana				33.7	33.5	33.2	33.1
Paraguay				17.6			moderate proportion of slum dwellers**
Peru	66.4	56.3	46.2	36.1	36.1		34.2
Suriname				3.9			7.3
Uruguay							no data ever recorded
Venezuela				32			high proportion of slum dwellers***

Notes:
*Chile No data available. UN MDG Progress Snapshot at country level records low proportion of slum dwellers, *https://mdgs.un.org/unsd/mdg/Resources/Static/Products/Progress2015/Snapshots/CHL.pdf*
**Paraguay No data available. UN MDG Progress Snapshot at country level records moderate proportion of slum dwellers, *https://mdgs.un.org/unsd/mdg/Resources/Static/Products/Progress2015/Snapshots/PRY.pdf*
***Venezuela No data available. UN MDG Progress Snapshot at country level records high proportion of slum dwellers, *https://mdgs.un.org/unsd/mdg/Resources/Static/Products/Progress2015/Snapshots/VEN.pdf*

Source: United Nations, Millennium Development Goals. Goal 7 target 7D: By 2020 to have achieved a significant improvement in the lives of at least 100 million slum dwellers

Data: Slum population as percentage of urban

slum upgrading programmes (i.e. improving sanitation and transport connectivity in precarious settlements). As these programs are mainly funded by the state, the growth of informality has serious consequences for local economies.

1.2 Economic drivers

Macro and micro economic conditions have a direct impact on the property market. Assuming perfect market conditions and a strong economy, rising employment rates can boost spending and the appetite for housing and property investment. Conversely, a struggling economy with high unemployment will have little surplus for spending among the population. These two scenarios combine in cycles where the economy moves through patterns of peaks and troughs of strengths and weaknesses. In this context, employment rates and consumer spending are useful economic indicators that can provide useful information on the purchasing power of individuals to buy, rent or invest in property.

South America follows economic cycles that are sometimes linked to global conditions and/or internal politics. For example, the poor performing economies of countries such as Brazil and Argentina (see Table 1.7), combined with the increased inequality and stagnation of poverty reduction measures claimed by populist governments, prompted the Congress in Brazil and voters in Argentina to support more centre-right parties and ideologies. In economic terms, this can be a radical change, as the new governments are now introducing policy reforms to adjust macroeconomic imbalances and improve the business environment. In theory, this should benefit the region's property market.

Table 1.7 GDP (total) for 2016 in billion USD at current prices

Argentina	541.748
Bolivia	35.699
Brazil	1,769.60
Chile	234.903
Colombia	274.135
Ecuador	99.118
Guyana	3.456
Paraguay	27.323
Peru	180.291
Suriname	4.137
Uruguay	54.374
Venezuela	333.715
Latin America	5,001.973
UK	2,649.89
US	18,561.93

Source: International Monetary Fund, World Economic Outlook Database, October 2016 (estimates)

Data: GDP (total) for 2016 in billion USD at current prices

At global level, the fall of commodity prices usually damages those countries that export raw materials. In the case of South America these are predominantly Argentina, Brazil, Colombia and Venezuela, some of the biggest economies in the region. As large international investors tend to follow general regional economic outlooks, the poor-performing GDP of these exporters of commodities can sometimes deter foreign investment. This affects the flow of cash coming into the region as a whole, therefore badly influencing South America's outlook in property markets, particularly large retail and office space transactions. Finally, corruption and institutional weakness can also damage the economy as this creates a climate of social unrest and mistrust in governments. The year 2016 saw a rise in protests due to corruption scandals in Brazil as well as in more stable countries such as Chile. As Chapter 6 will explain, corruption and weak institutions are another factor that can deter direct foreign investment into real estate.

1.2.1 Economic indicators

The most important variables to understand an economy are GDP and employment data. After presenting these variables, this section will present other information such as interest rates, inflation, and bank reserves with the aim of illustrating the depths of the capital markets and the availability of credit in South America. Most data collected here is from reliable open sources including World Bank and IMF; forecasts by these and other expert agencies have also been included in order to give a medium to long-term view of the economies. However, data collected by these institutions rely on countries' official reports, which are sometimes not submitted on time using controversial methods for calculations.[3]

To correct potential errors, data from different sources was collected and compared to find correlations. Additionally, estimates and forecast should be taken with caution and are given for illustration purposes only. As in the previous section, data from the UK and the US has also been collected as well as Latin American countries in order to provide regional and global context to the data presented here. Finally, website links to all databases are also included in the tables, as they can provide further information on the methodology of databases and forecasts, which cannot be included in the present volume due to spatial constraints.

i) GDP

In terms of economic outputs, South America has a combined GDP of USD 3,558.50 billion compared to a total of USD 5,001.973 billion for Latin America as a whole (Table 1.7). It can certainly be argued that the Southern part of the region is its economic driver, producing nearly 70% of Latin America's total outputs. Notwithstanding, even within this subgroup, there are some countries that are clearly dominant, including, by order of productivity: Brazil, Argentina, Venezuela, Colombia, Chile and Peru. Ecuador's outputs are nearly half of Peru's while the rest of the countries (Bolivia, Guyana, Paraguay, Suriname and Uruguay) are very small players.

In terms of GDP growth (Table 1.8), South America has been on a downward trend since the commodity prices started to decline in 2011, which explains the

Table 1.8 GDP growth (%) forecast

	2016 (estimate)	2017	2018	2019
Argentina	−1.761	2.731	2.767	2.897
Bolivia	3.7	3.9	3.5	3.5
Brazil	−3.273	0.2*	1.5*	1.967
Chile	1.7	1.986	2.7	3.02
Colombia	2.171	2.689	3.8	4.3
Ecuador	−2.265	−2.699	−1.114	−0.419
Guyana	4.025	4.077	3.934	3.996
Paraguay	3.507	3.64	3.785	3.91
Peru	3.749	4.116	3.578	3.527
Suriname	−6.976	0.5	1.052	2.3
Uruguay	0.1	1.2	3.1	3.4
Venezuela	−10	−4.5	−3	−1
UK	1.837	1.5*	1.4*	1.792
US	1.578	2.3*	2.5*	1.924

Notes:
* Data updated January 2017

Source: The World Bank, World Development Indicators (2016). GDP growth (annual %) [Data file]. Retrieved from http://data.worldbank.org/indicator/NY.GDP.MKTP.KD.ZG

Data: GDP growth (% annual) for the last ten years up to 2016

generally modest estimated growth rates provided by the IMF for most of the countries. The negative rates for Argentina, Brazil, Ecuador, Suriname and Venezuela can also be attributed to a mix of political changes and corruption scandals (see Introduction for politics and Chapter 6 on corruption).

As a result, the prospects for Brazil, the regional giant, are modest if compared, for example, with the +6% projections for China and India given by the same source (IMF) for the years 2017 and 2018. The best regional growth in South America is expected in Colombia and Uruguay, with each one nearly doubling current estimated outputs by 2019. Argentina, Chile and Paraguay will also experience growth, while Bolivia is expected to maintain current levels of outputs. Guyana and Peru's economies are both contracting, although their outputs are still amongst the regional best. Ecuador and Venezuela are the poorest performers with the latter firmly on the lead.

ii) Unemployment

Unemployment rate is a lagging indicator, which means it has a delayed reaction to changes in the economic cycle. This is because in the case of a recession, it takes time for firms to respond to loss in trade and dismiss the surplus workforce as employees are usually protected by contracts. Conversely if the economy expands, it also takes time for employers to find and hire employees. It is important to understand the lagging qualities of this indicator as it has a direct impact on the demand and supply of built space for all sectors, but most particularly for office space.

As Table 1.9 indicates, unemployment rates across the countries vary greatly. Furthermore, some countries tend to disregard hidden unemployment from the national statistics – this means they only consider those who are actively seeking work as unemployed. According to the International Labour Organisation (ILO), these countries include Colombia and Ecuador. In order to correct potential errors, data from two different sources have been crosschecked to assemble Table 1.9, representing estimates from ILO (marked with *) and IMF sources. There is no available data for Guyana, while data available for Bolivia and Peru must be considered carefully and with some scepticism. Chapter 6 looks in more detail at the lack of available data in the region and what effect this has to RE investment.

The ILO has reported that the labour situation in Latin America as a whole has worsened due to the economic slowdown (ILO, 2016). The report highlights that the average unemployment rate increased from 6.6% in 2015 to 8.1% in 2016. Undoubtedly, the political situation in Venezuela is driving the average upwards, while the long-term prospect in Brazil looks set in the double digits for the next four years. For comparison, the normal unemployment rate in the US is around 5% according to some authors (Weidner and Williams, 2011) while the recorded rate for the European Union for February 2017 was 8.0%.[4] IMF's

Table 1.9 Unemployment rate

Country	2016	2017	2018	2019	2020	2021
Argentina	9.3*	8.466	8.331	7.491	6.949	6.769
Bolivia	4	4	4	4	4	4
Brazil	11.3*	11.539	11.076	10.441	10.045	9.995
Chile	6.6*	7.638	7.2	7	6.6	6.3
Colombia	9.6*	9.6	9	8.9	8.8	8.7
Ecuador	5.4*	6.899	6.899	6.697	6.858	6.294
Guyana	n/a	n/a	n/a	n/a	n/a	n/a
Paraguay	8.3*	5.474	5.492	5.239	5.141	5.06
Peru	4.4*	6	6	6	6	6
Suriname	11.879	11.692	11.263	10.396	9.357	8.204
Uruguay	8*	8.512	8.183	7.722	7.332	7.184
Venezuela	18.148	21.378	23.96	26.928	28.361	29.766
UK	4.962	5.201	5.448	5.5	5.425	5.425
US	4.895	4.771	4.729	4.846	5.049	5.055

* **ILO** based on official data from household surveys of the countries

Source: International Monetary Fund, World Economic Outlook Database, October 2016, and International Labour Organisation Latin American and Caribbean Labour Overview 2016

Data: Unemployment (%) of total labour force

forecast for Argentina, Chile and Paraguay looks promising, while the rest of the countries are predicted to remain above 7% until 2021.

In the case of Paraguay the positive outlook could be down to just one city: Ciudad del Este, which is located near the Brazilian border and offering cheap labour for manufacturing jobs (as well as government tax incentives and lower energy prices). As a result, Brazilian companies are relocating to Ciudad del Este and some of them even moving operations from China to Paraguay.[5] This can explain the dramatic lowering of the unemployment rate in this country.

An additional positive outlook for the other countries can be inferred from the demographics. As stated before Argentina, Brazil and Chile are the ones absorbing the most regional migrants, and the gradual lowering of unemployment seems to indicate that these economies are coping well with this influx. However, it is worth pointing out that unemployment fluctuations in the region can be lessened by the informal sector as workers usually find other means of employment. In this regard, a comparative chart with the size of the informal economy of the countries under study here can also provide an indication of the state of the labour market. Table 1.10 presents an estimated size of the informal economy; data collected is the average of two main sources: an IMF paper (Vuletin, 2008), which has data corresponding to the years 2000 to 2007; and ILO data covering the years 2009 to 2013. The two sources have helped to fill data gaps but the estimated final average should be taken for illustration purposes only.[6]

The final column of the table classifies the countries according to OECD ranges of informality across the world, which are: below 15%; between 15–30%; between 30–45%; and more than 45%.

Table 1.10 shows that in South America there are two main OECD ranges of informality: between 30–45%, which includes Argentina, Brazil, Chile, Guyana

Table 1.10 Informal economy

	% for 2008	%	year	2017 (average)	OECD categories (CEPLAN, 2016)
Argentina	32.9	n/a	n/a	32.9	30–45%
Bolivia	n/a	71.8	2009	71.8	above 45%
Brazil	28.4	36.9	2013	32.65	30–45%
Chile	32.1	n/a	n/a	32.1	30–45%
Colombia	43.5	60.2	2014	51.85	above 45%
Ecuador	50.7	40.4	2015	45.55	above 45%
Guyana	36.7	n/a	n/a	36.7	30–45%
Paraguay	68.2	64.5	2013	66.35	above 45%
Peru	38.1	74.3	2012	56.2	above 45%
Suriname	n/a	n/a	n/a	n/a	n/a
Uruguay	36.2	24.7	2015	30.45	30–45%
Venezuela	43	47.5	2009	45.25	above 45%

Source: Vuletin (2008) IMF working papers ILO, www.elibrary.imf.org/view/IMF001/09218-978145 1869637/09218-9781451869637/09218-9781451869637.xml?redirect=true

Data: Estimated size of informal economy

and Uruguay; and above 45% corresponding to Bolivia, Colombia, Ecuador, Paraguay, Peru and Venezuela.

In countries with high rates of population growth and urbanisation, the informal sector tends to absorb most of the expanding labour force in the urban areas (Benjamin et al., 2014). In South America there seems to be a correlation between countries that fall in the 15–30% range of economic informality and high levels of urbanisation (Table 1.5); while most countries in the >45% range are below the 80% urban threshold. The exception is Venezuela, which has 88.9% of population living in cities while simultaneously having over 45% of economic informality; still, this singularity can be explained by the current political and economic crisis that the country is facing (see Introduction).

A final comparison can be drawn between the level of economic informality and the spatial informality. Table 1.11 presents a comparative analysis of Tables 1.10, 1.5 and 1.6, grouping countries according to the two ranges of economic informality found in the region (between 30–45% and >45%), while simultaneously ranking them according to percentage of urbanisation (Table 1.5), and population living in slums (Table 1.6).

The outlier country in the 30–45% group is Guyana, which is the least urban, as all others in this group are above 80%. Guyana also has the highest percentage of population living in slums within the group. There are similarly interesting exceptions in the >45% informality group of countries. The most notable is Colombia, which has room for urban growth and a low percentage of slum dwellers considering the overall regional context. A positive interpretation of

Table 1.11 Comparative analysis of informality and urbanisation

Source	Table 1.10	Table 1.5	Table 1.6
	OECD categories (CEPLAN, 2016)	Percentage urban	% of population living in slums
Uruguay	30–45%	95.2	no data ever recorded
Argentina	30–45%	91.6	16.7
Chile	30–45%	89.4	low proportion of slum dwellers
Brazil	30–45%	85.4	22.3
Guyana	30–45%	28.5	33.1
Venezuela	above 45%	88.9	high proportion of slum dwellers
Peru	above 45%	78.3	34.2
Colombia	above 45%	76.2	13.1
Bolivia	above 45%	68.1	43.5
Ecuador	above 45%	63.5	36
Paraguay	above 45%	59.4	moderate proportion of slum dwellers
Suriname	n/a	66.1	7.3

Data: Comparative analysis

the data available for Paraguay in terms of slum population, can also align this country's perspectives to Colombia's. Finally, Peru, Bolivia and Ecuador will need to implement new ways of controlling the informal economy to avoid reaching Venezuela-like conditions of high urbanisation, high informal economy and high levels of population living in slums.

These general observations can help to narrow the focus of study in the present book, and concentrate on the strongest economies that can support an analysis of real estate fundamentals.

1.3 Selection of strongest economies

Table 1.12 groups and ranks the countries according to four variables: population in 2016 (Table 1.2); GDP 2016 (Table 1.7) and forecasts for same set of data for 2030 (Table 1.4) and 2019 (Table 1.8), respectively. The higher the ranking position obtained by each country, the lower the points attributed to them in the scale from 1–12. Table 1.13 presents the results of the final scoring; the lower the score means the country has ranked at the top more consistently, so a low score means a strong economy with a good forecast.

The surprising result is that Colombia is at the top of the list, mainly because Brazil has been given a very low GDP forecast by the IMF (Table 1.8) as has Argentina. Still, it is Colombia's demographics that are helping it to obtain good results rather than a strong GDP. On the contrary, a combination of low demographics and low forecasted GDP has placed Chile below Peru. This is another surprise as in GDP terms alone, Chile usually ranks higher than Peru. The predictable scenario is the decline of Venezuela, which was until not long

Table 1.12 Ranking of economies by size

	Ranking of countries		Ranking of countries			Ranking of countries		Ranking of countries	
	Table 1.2 Population		Table 1.7 GDP			Table 1.4 Population forecast		Table 1.8 GDP forecast	
	2016		2016			2030		2019	
1	Brazil	206.1	Brazil	1,769.60		Brazil	223.1	Colombia	4.3
2	Colombia	48.8	Argentina	541.748		Colombia	53.3	Guyana	3.996
3	Argentina	43.6	Venezuela	333.715		Argentina	49.1	Paraguay	3.91
4	Peru	31.5	Colombia	274.135		Venezuela	36.1	Peru	3.527
5	Venezuela	31	Chile	234.903		Peru	35.9	Bolivia	3.5
6	Chile	18.2	Peru	180.291		Chile	20.3	Uruguay	3.4
7	Ecuador	16.5	Ecuador	99.118		Ecuador	19.7	Chile	3.02
8	Bolivia	11	Uruguay	54.374		Bolivia	13.3	Argentina	2.897
9	Paraguay	7	Bolivia	35.699		Paraguay	8.2	Suriname	2.3
10	Uruguay	3.5	Paraguay	27.323		Uruguay	3.6	Brazil	1.967
11	Guyana	0.8	Suriname	4.137		Guyana	0.8	Ecuador	–0.419
12	Suriname	0.5	Guyana	3.456		Suriname	0.6	Venezuela	–1

Table 1.13 Ranking results

Country	Points
Colombia	9
Brazil	13
Argentina	16
Peru	19
Venezuela	24
Chile	24
Uruguay	29
Bolivia	30
Ecuador	32
Paraguay	35
Guyana	36
Suriname	44

ago the third strongest economic power in South America in terms of outputs. As explained in the introduction, the political instability of the country is too substantial and there are no signs of change or recovery in the short or medium term. This fact, combined with the poor quality of urbanisation explained in the previous section, has meant that Venezuela will not be included in this study. Equally, the countries that have scored above 25 points in Table 1.13 (ranking), will not be the focus of this book. Their economies are just too small and in some cases the availability of robust datasets are very scant.

The remainder of the volume will therefore focus on the top five countries: Colombia, Brazil, Argentina, Peru and Chile, which are named hereafter as the "Big Five." As the next section will demonstrate, they are all at different stages of economic development; therefore, the sophistication of the real estate investment tools vary greatly from country to country. Still, they are all actively taking steps towards converging via trade agreements. Brazil and Argentina are traditionally the longest trading partners, related via the Mercosur, while Colombia, Chile and Peru are connected via the Pacific Alliance.

1.4 Real estate market fundamentals for the Big Five

Having reviewed the key economic drivers for South American countries, this next section now looks more closely at those indicators that affect property investment decisions, such as interest rates and inflation. It also examines the depth of the markets and the availability of credit for mortgages and other RE operations.

1.4.1 Interest rates (IR) and inflation

The impact of IR on property manifests in different ways. Assuming perfect market conditions, low IR means low borrowing costs to buy homes, which can in turn increase the demand for properties. For developers, if the borrowing costs are low, they are able to satisfy the higher demand while making a profit. Conversely, if IR are high, it means high borrowing costs, which affect the required rate of return for developers and potentially reduce their ability to supply new units. Equally, high borrowing costs affect the demand as buyers are constrained from borrowing to purchase properties. A RE market can become stagnant in a context of high IR as any activity is rendered impossible for both sides, buyers and developers. Equally, business and entrepreneurial activities can suffer as properties cannot be used as collateral to raise capital for new enterprises.

In simple economic terms, inflation occurs when the price of goods escalates, because people's purchasing power begins to fall and more money is needed to buy the same goods. To control this, IR are increased to slow the demand for goods, which in turn reduces the prices. Equally when the economy slows, IR are lowered to allow for more demand of goods and services in order to avoid deflation. Since controlling inflation impacts interests rates and interest rates impact property investment, inflation is also an indicator of the health of a real estate market.

Information for IR for the Big Five are provided in Table 1.14 while Tables 1.15a and 1.15b present historical and current inflation rates for all countries.

Most countries attempt to control inflation by imposing a target, which, as seen in Table 1.15b, varies for the different economies. In our selection of countries, the UK and US follow a rigid 2% while South American countries have more flexibility with a variance of +/− 1 percentage point. Argentina is the odd one out with an 8-point percentage change. With the exception of Brazil and Chile, all other countries including the advanced economies are missing the targets set by the corresponding central banks of each country (Table 1.15a). In the case of Argentina, the country's high interest rates are hardening the control of

Table 1.14 Interest rates

	Date of reporting	*2016 (previous)*	*2017 (current)*	*2018 (forecast)*
Argentina	April 2017		26.25	
Brazil	April 2017	13.75	11.25	13.25
Chile	April 2017	3	2.75	2.75
Colombia	March 2017	7.25	7	7
Peru	April 2017	4.25	4.25	4.25
UK	March 2017	0.5	0.25	0.25
US	March 2017	0.5	0.75	1

Source: International Monetary Fund, International Financial Statistics (IFS) and World Bank. Trading Economics, www.tradingeconomics.com/country-list/interest-rate

Data: Current, previous and forecast IR with date of current reporting by central bank. Data on current IR has been cross-checked with IMF and World Bank databases for accuracy. All sources report same data including the omission of Argentina's historical data. Forecasts by Trading Economics using its own methodology.

Table 1.15a Inflation rates

Country Name	2006	2007	2008	2009	2010	2011	2012	2013	2014	2015	2016
Argentina	10.90	8.83	8.58	6.28	10.78	9.47	10.03	10.62	n/a*	n/a	35**
Brazil	4.18	3.64	5.66	4.89	5.04	6.64	5.40	6.20	6.33	9.03	8.74
Chile	3.39	4.41	8.72	0.07	1.41	3.34	3.01	1.79	4.40	4.35	3.79
Colombia	4.30	5.54	7.00	4.20	2.28	3.41	3.18	2.02	2.88	5.01	7.52
Peru	2.00	1.78	5.79	2.94	1.53	3.37	3.65	2.82	3.23	3.56	3.60
UK	2.33	2.32	3.61	2.17	3.29	4.48	2.82	2.55	1.46	0.05	0.64
US	3.23	2.85	3.84	−0.36	1.64	3.16	2.07	1.46	1.62	0.12	1.26

Notes:
* National Institute of Statistics and Census revised national accounts from 2004–2015. Argentina, which was temporarily unclassified in July 2016 pending release of revised national accounts statistics, is classified as upper middle income for FY17 as of September 29, 2016.

The World Bank systematically assesses the appropriateness of official exchange rates as conversion factors. An alternative conversion factor is used when the official exchange rate is judged to diverge by an exceptionally large margin from the rate effectively applied to domestic transactions of foreign currencies and traded products. In the case of Argentina, the World Bank has found that during 2012–2015 there were two exchange rates (official and parallel) and parallel exchange rate (blue chip swap rate) was used in around 20% of the transactions. Based on this information an alternative conversion factor has been calculated using a weighted average method for this period.
** as reported by Trading Economics

Source: World Bank Development Indicators for data from 2006–2016. Trading Economics for 2017 including data of official reporting by respective governments, www.tradingeconomics.com/country-list/inflation-rate

Data: Inflation, consumer prices (annual %) – inflation as measured by the consumer price index reflects the annual percentage change in the cost to the average consumer of acquiring a basket of goods and services that may be fixed or changed at specified intervals, such as yearly

Table 1.15b Inflation targets

Country name	2017	Date	Bank name	Target
Argentina	40.5	April 2017	Central Bank of Argentina	12–17 +8
Brazil	4.57	March 2017	Central Bank of Brazil	4.50 +/– 1.5
Chile	2.7	March 2017	Central Bank of Chile	3 +/– 1
Colombia	4.69	March 2017	Central Bank of Chile	3 +/– 1
Peru	3.97	March 2017	Central Reserve Bank of Peru	2 +/– 1
UK	2.3	March 2017	Bank of England	2
US	2.4	March 2017	Federal Reserve Bank	2

Source: Trading Economics, www.tradingeconomics.com/country-list/inflation-rate; and target data from Central Banks News, www.centralbanknews.info/p/inflation-targets.html

Data: Inflation current and target (annual %) as reported by central banks including month of reporting the inflation. In Brazil, inflation rate slowed in 2017 to 4.57% after peaking in 2015 at 9.03%. This has allowed for a reduction in IR from 13.75% to 11.25%. In Colombia the inflation rate eased from the previous year as costs for food and housing rose less than previous years. As a result, IR were lowered by 0.25% by the Central Bank of Colombia. Peru's constant upward inflation trend is due to heavy flooding, which has had an immediate impact on crops and food prices. In Chile, the upward trend in 2016 can also be explained by earthquakes and tsunamis that have affected the country in previous years, particularly 2015, which is one of the highest inflation years for this economy in recent times. Recovery from such events take time but the Central Bank of Chile has managed to reduce IR two years after the last earthquake that affected the country, which shows its resilience.

1.4.2 Wage levels

In a strong economy there is more activity and more positions to fill; therefore, unemployment levels are down and wages are boosted given that labour is in demand. Higher wages usually mean higher levels of disposable income and greater opportunities for spending and for property investment. Notwithstanding, income distributions must also be taken into consideration particularly for highly unequal societies such as in Latin America. Indeed a World Bank report (2013), maintains that despite the rise of the middle classes that happened during the region's boom years of 2009–11, around 80 million people still live in extreme poverty, a quarter of them are in Brazil; and with a further 40% at risk of falling back into poverty if the recession continues to hit or if extreme weather events continue as a result of climate change.

The previous section just demonstrated how inflation caused by natural disasters can affect central bank's decisions on IR, as in the cases of Peru and Chile, and how IR affects property investment and hinders the collateralisation of property for entrepreneurial activities, which in turn can have a detrimental effect on productivity and economic growth. But the links between inequality and economic growth must also be taken into account in a region with low distribution of wealth. The subject has been widely researched, with the most recent contributions from French economist Thomas Pickety, who has demonstrated how wealth around the world has concentrated in fewer hands over the years; he proposed in his book *Capital* (2013) a system of progressive taxation at the global level to prevent the deepening of wealth concentration. On the other hand, British scientists Wilkinson and Pickett (2009) argue that inequality affects all aspects of society, from people's mental health to life expectancy, and that the collective unhappiness of the inhabitants of an unequal society affects the economic stability of countries. Finally, the American economist Joseph Stieglitz (2013) has also argued that unhealthy gaps of wage levels and inequality can slow economic activity and GDP. All these authors from varying countries agree that a meaningful minimum wage can help societies to reduce poverty and with it, lessen the income distribution gap in order to boost living conditions and GDP.

Table 1.16 compiles minimum wages across all countries as reported by their governments in March 2017 and compiled by Trading Economics. Table 1.17

Table 1.16 Average minimum wages

	Current (March '17)	Previous	Increase
Argentina	AR$ 8,062	AR$ 7,560 (October '16)	6.2%
Brazil	BRL 937	BRL 880 (October '16)	6.08%
Chile	CLP 265,000	CLP 250,000 (October '16)	5.6%
Colombia	COP 737,717	COP 686,077 (October '16)	7%
Peru	PEN 850	PEN 750 (April '16)	11.76%
UK	£1,440*	£1,382.4 (March '16)	4%
US	USD 1,392**	same as 2016	0%

Notes:
* UK wages based on a £7.50 hourly rate for a person working 48 hours per week for 2017 and £7.25 for March 2016. Same source with authors' calculations. For March 2016 UK Government press release, www.gov.uk/government/news/the-government-accepts-minimum-wage-rate-recommendations – 2
** US wages based on a £7.25 hourly rate for a person working 48 hours per week same source with authors' calculations

Source: Trading Economics, www.tradingeconomics.com/country-list/interest-rate

Data: Real minimum monthly wages in local currency

Table 1.17 Cost of living

	McMeal at McDonald's (or equivalent combo meal)	Petrol (1 L)	One-way ticket, local transport	Basic utilities (electricity, heating, water, waste collection) for a 85m2 apartment	Rent (1 bedroom apartment in city centre)
Argentina	AR$ 125	AR$ 19.12	AR$ 7	AR$ 1,369.34	AR$ 6,257
Brazil	BRL 25	BRL 3.70	BRL 3.80	BRL 240.22	BRL 1,307.47
Chile	CLP 3,800	CLP 750.28	CLP 670	CLP 75,650.53	CLP 273,361.73
Colombia	COP 15,000	COP 2,156.95	COP 2,000	209,637.61	831,448.82
Peru	PEN 15	PEN 3.51	PEN 1.50	PEN 141.64	PEN 1,784.91
UK	£5.00	£1.12	£2.40	£144.57	£798.37
US	USD 7	USD 0.61	USD 2.25	USD 148.9	USD 1,164.89

Source: *numbeo.com* data based on thousands of monthly contributions from all cities in each country. Data updated monthly – April 2017

Data: Illustration of cost of living based on prices of meals, transport and accommodation

intends to build a comparative picture of the cost of living using latest data available at Numbeo (see table details for their methodology of data collection). Included in the latter is the cost of a meal, which has been largely recognised as a proxy for international comparisons.[7] Also in the table is the cost of transport as well as rent and utilities. All costs are given in the local currency and represent the country average. This means that the samples of rental values, for example, are not related to any particular location but are a country average, which includes expensive capital cities as well as cheaper locations. There are of course some errors and biases with the data presented but the goal here is not to test a new theory, it is rather to offer the reader a relatable proxy of the cost of living in relation to the minimum wages in each country.

As seen from Tables 1.16 and 1.17, one person paying rent and utilities for a small apartment (82m²) is left with very little money to survive in Argentina, as this expense signifies 94% of the total wage. For the rest of the countries, the option is unsustainable as these outgoings exceed the minimum wage. Considering that there are two contributors to a household, then the accommodation expense including rent and utilities will take up to 65–70% of the combined wages; leaving 30% of the minimum wage to cover for other living expenses. The exception is Peru, which will take three minimum wages to be at the same level of the other countries.

For comparison, data from the UK and the US have also been collected.[8] The figures illustrate that the US has a similar problem as the rest of South America, given that after paying all outgoings for the accommodation, a person is left with only USD 5.62 to afford all other monthly living expenses. The UK has a better standard as accommodation represents around 65% of all monthly expenses of one person's wages.

Still, it can be argued that rental levels and the quality of housing varies considerably from one country to the next, so we can further test the effectiveness of minimum wages by using the burgernomics approach (see endnote vii). As a proxy for purchasing power parity, Table 1.18 shows how many burgers a minimum wage can buy in all countries. Also calculated are the cost of transportation based on car usage and on public transport.

Based on the percentages presented in Table 1.18 for a person living on the local minimum wage, the US seems the most affordable country, followed by the UK. Argentina and Chile are the most affordable amongst the South American countries, mainly thanks to their low public transport costs and the affordability

Table 1.18 Cost-of-living comparison

Source	*Table 1.17*		
	Burgernomics	*Percentage of filling the tank to salary (based on weekly fill of a 32 L car) %*	*Percentage of monthly commuting to salary (based on one return ticket ×5 times/week) %*
Argentina	64.00	30	3.00
Brazil	37.00	50	16.00
Chile	70.00	36	10.00
Colombia	49.00	37	11.00
Peru	57.00	53	7.00
UK	288.00	10	7.00
US	197.00	6	6.00

Data: Authors' calculations

of food (as in quantity of burgers). Peru is the most expensive in terms of petrol, but has good levels of public transport cost. Colombia and Brazil are the most expensive countries, consistently having higher costs related to transport, fuel and food than the others. The fact that the UK and the US are more affordable than the South American countries is no surprise considering that wage levels are higher in those countries in relation of their own cost of living. Additionally, tax evasion in Latin America is a widespread problem (Gómez Sabaini and Jiménez, 2012) and the only resource most governments have to collect revenues is through taxing the price of goods, which is another illustration of how badly designed policies are affecting the most vulnerable in the region, as the poor are bearing the brunt of the taxation system via added value tax on small (but visible) items. Another explanation for the higher cost of food can be that the exchange value of currencies is distorted, or that the labour/facilities/transportation of goods in the production chain are more expensive. Notwithstanding, the unsustainable level of minimum wages is evident, and in the context of widespread regional inequality, and spatial and economic informality, the current policies for all countries appear ill suited.

Furthermore, some countries' inflation levels are worrying. In Argentina the wage increase from October 2016 to March 2017 of 6.2% (Table 1.16) is dwarfed against a 40% inflation (Table 1.15). Although the country is number one in the global top 20 Real Wage Increase Forecast Ranking produced by ECA International,[9] the higher wages have been mainly driven by the inflation rather than an increase in productivity. This fact leaves Chile as the country that is performing best in terms of effectiveness of minimum wages and living costs. It is also the country where the price of the McMeal at McDonald's (or equivalent combo meal) is lower than in the developed economies (USD 5.7 at an exchange rate of 0.0015 for April 2017).

A final test of the effectiveness of minimum wages can be provided by the Gini coefficient. This is a measure of income distribution of a country's population and ranges from 0 to 1, the closest to 0 the more equality and vice versa, the closest to 1 the more inequality. As reference the US is considered an unequal society with a Gini coefficient of 0.4106, while one of the most equal societies is Norway with 0.259.[10] Amongst the South American countries, wealth distribution varies with Argentina showing a Gini coefficient of 0.427; Brazil, 0.515; Colombia, 0.535; Peru, 0.441 for 2014; and Chile 0.5045 for 2013 (World Bank estimates). Argentina and Peru are the most equalitarian countries followed by Chile. Again Brazil and Colombia are poor regional performers with high levels of inequality, which confirms the previous results shown thus far as the most expensive countries in the region for those earning a minimum wage.

1.4.3 Consumer spending

Consumer spending is a real-time indication of the health of the economy. Unlike employment, which can take time to adjust to economic cycles and changes, the availability of spending money in households identifies a much faster reaction from the public.

Table 1.19 presents World Bank data for consumer spending for all countries under study between 2013 and 2015, as well as the UK and the US for comparison. While Brazil is the only country reaching the trillion mark set by the two advanced economies, all other countries are in the billions. Note that Argentina is in a clear second place, ahead of Colombia despite the latter's population being 5.2 million more than the former country (Table 1.2). Undoubtedly, Chile, with the smallest population, is the country with the most consumption in South America, while Peru is the least.

The historical data shows that Chile, Colombia and Peru are on an upward trend for consumer spending, while Argentina and Brazil are stable. The impact of this indicator on the development of shopping centres will be seen in Chapter 2.

1.4.4 Monetary policies and bank reserves

South American countries with their high reliance on commodities have not only been hit by international price reductions in raw materials but also by a depreciation of their currencies. As an illustration from 2014, the Brazilian Real has declined by more than 40% against the US dollar, the Colombian Peso 35%; Chile 20% and Peru 15%. The Argentinean Peso received a nominal devaluation of 40% after the political changes in 2015 that released the exchange control mechanisms (Powell, 2016). Some of the currencies' downward trends are the largest and most sustained episodes since the early 1980s (Brazil and Colombia, for example) and the IMF forecasts that the downward pressure on exchange rates will continue for some time due to slow global growth and low levels of commodity prices (IMF, 2016b).

Bank reserve levels vary greatly and there is no agreement in what should be the recommended one (Rule, 2015). Still, low levels can affect the percentage

Table 1.19 Consumer spending

Country/year	2013	2014	2015
Argentina	308,782,963,669	292,743,406,455	307,454,151,645
Brazil	1,492,241,366,942	1,511,952,313,799	1,451,515,559,624
Chile	156,374,788,690	160,088,373,016	163,114,740,072
Colombia	206,071,932,175	214,628,701,992	223,098,243,544
Peru	110,879,641,971	115,227,234,013	119,134,460,795
UK	1,629,152,308,096	1,664,338,286,774	1,706,039,180,470
US	10,740,008,937,800	11,048,479,099,500	11,400,068,298,000

Source: The World Bank, World Development Indicators (2016). GDP growth (annual %) [Data file]. Retrieved from http://data.worldbank.org/indicator/NE.CON.PRVT.KD

Data: Household final consumption expenditure (constant 2010 US$)

for investment (such as mortgages), which can in turn affect interest rates. For example, governments with higher reserves and wanting to increase credit might opt to reduce interest rates to boost more property transactions. Table 1.20 shows bank reserves across all countries under study; Bolivia has been added as a regional example of a country with very little reserves while the UK and the US are also given as comparison.

Most countries are increasing their bank reserves from 2016 to February 2017, with the exception of Chile. The new Argentinean government issued a tax incentive programme that intended to repatriate Argentineans' cash deposited abroad; the programme concluded in December 2016 and part of the increase in central bank reserves (approximately USD 5,300 million) has been attributed to the success of this measure.[11] It remains to be seen if the new deposits can translate now to an increase in loan offers from the banks.

1.4.5 Capital markets

About 35% of emerging markets' corporate bonds from around the globe are issued partially or wholly by government-owned companies (IMF, 2009). This fact is particularly relevant in Latin America and within the oil sector where the ratio is close to 70%. In South America this includes Argentina's YPF, Brazil's Petrobras and Chile's Codelco. This factor highlights the fact that contrary to Europe and other developing nations, where most of the debt is concentrated in the private sector, South America's debt is owned by governments. This exposes sovereign balance sheets to additional risks, including the international volatility of oil prices and currency fluctuations.

Table 1.20 Bank reserves

	2016*	2017 M2	2017 M2 Gold (including gold deposits and, if appropriate, gold swapped)
Argentina	36,323	50,608	2,291.49
Bolivia	8,487	10,353.12**	1,638.06**
Brazil	362,505	370,111.12	2,969.19
Chile	40,484	39,709.91	9.98
Colombia	45,962	46,442.55	229.98
Peru	61,594***	62,250	1,390
UK	123,500	162,951	12,533
US	106,291	116,244.60	11,041.06

Notes:
* Data does not include gold reserves
** Data corresponding to 2016 M11
*** Data from World Bank for Peru is for 2015 and includes gold reserves for the year

Source: International Monetary Fund, International Financial Statistics (IFS) and World Bank

Data: Bank reserves for 2016 (total reserves US$ million)

Given this scenario, several South American countries such as Peru and more recently Argentina are interested in increasing corporate bonds issuing by the private sector as well as attracting cross border investment by pension funds and insurance companies. This interest is mainly driven by the need to fund large infrastructure and social housing projects via public-private partnerships (PPP) with the aim of diminishing the strain on the state (see Chapters 2 and 5). Such an increase in cross-border investment would need to be accompanied by an appropriate risk management strategy in order to attract investment. Additionally, capital markets need to have sufficient depth to guarantee liquidity for investors, and increase accessibility to a wide range of domestic and international investors at a low transaction cost.

Leverage across emerging markets has increased in recent years as state-owned enterprises and local firms have managed to access international capital markets (Powell, 2016). In this context South America is no exception and is making a great leap forward in improving the level of sophistication of its capital markets. For example, the introduction of hybrid securities (combining both debt and equity characteristics) in Colombia[12] or the issuing of green investment funds with a sustainable mandate in Brazil.[13] Equally, non-financial firms in countries such as Brazil, Colombia and Chile have issued bonds and the outstanding stock within this sector has tripled from 2008 until 2016. In sum, the combined financial activities of state-owned and corporate companies are substantially increasing the depths of capital markets in South America.

From all the countries under study here, Argentina is the one that is still suffering with high inflation, which in turn restricts availability of long-term financing products. However, the country has made a comeback in the international debt market after arranging a deal with holdout creditors in the first quarter of 2016, and then issuing USD 16.5 billion four-tranche bond deal. The return of one of the largest economies in South America has certainly been behind the rise of the regional sovereign bond issuing manifested in 2016 as reported by Latin Finance.[14] Simultaneously, some bonds have been issued in local currency (Chile and Brazil), which indicates a deepening of their local markets thanks to an improved public debt management strategy (Arslanalp and Tsuda, 2014; see also Chapter 5 in this volume). In advanced economies with more sophisticated banking systems, debt financing mainly happens via local currency loans, which offers other mechanisms for risk collateralisation and diversification. The move to a local currency bond issuing by Brazil and Chile indicates that these two countries' capital markets are at the forefront of South America (see Chapter 4).

In general the regional rise in the bond-type of financing shows the greater integration of the region to the global economy. As an illustration, Table 1.21 shows that central government's debt as percentage of GDP is increasing in most countries, with the exception of Colombia and Peru.

According to Forbes, the most indebted economies in the world include Japan, with a debt of 227.2% of GDP followed by Greece, with 175,1%. In sixth place

Table 1.21 Government debt to GDP ratio

	2007	2008	2009	2010	2011	2012	2013	2014	2015	2016
Argentina*	44.40	39.20	39.60	36.10	33.30	37.30	40.20	48.60	48.40	n/a
Brazil	62.45	60.66	63.81	61.32	59.34	59.85	57.23	58.46	67.48	69.49
CHILE*	3.90	4.90	5.80	8.60	11.20	11.90	12.80	15.10	17.50	n/a
Colombia	32.7*	64.76	68.58	72.34	62.77	65.45	58.63	37.9**	42.5**	41**
Peru	28.53	26.08	26.40	23.68	20.72	19.22	18.33	19.32	22.65	n/a

Notes:
* Trading Economics data
** Banco Central República de Colombia provisional figures

Source: World Bank World Development Indicators, for Argentina and Chile Trading Economics www.tradingeconomics.com/country-list/government-debt-to-gdp for Colombia Central Bank of Colombia www.banrep.gov.co/economia/pli/bdeudax_t.pdf

Data: Central government debt, total (% of GDP) Debt is the entire stock of direct government fixed-term contractual obligations to others outstanding on a particular date. It includes domestic and foreign liabilities such as currency and money deposits, securities other than shares, and loans. It is the gross amount of government liabilities recuced by the amount of equity and financial derivatives held by the government. Because debt is a stock rather than a flow, it is measured as of a given date, usually the last day of the fiscal year.

is the US with a debt of 101.5% of the national GDP.[15] Borrowing is a decision made by central governments; for example, when the economy is slowing, governments might borrow capital to invest in national projects (infrastructure, for example) that can increase jobs and pump-up activity. Some authors suggest that the debt to GDP ratio should be kept below 70–80% (Checherita-Westphal and Rother, 2012). However, it is worth mentioning that higher levels of debt are no indication of the degree of development of a country, and an important question is who owns the debt. For example, Japan's debt is owned mostly by domestic investors, so the government is less exposed to foreign investors and there is less risk of capital flight in case of a downturn in investor's confidence. As Checherita-Westphal and Rother (2012) explain, this is not the case for many European countries such as Greece, where most of its debt is in foreign hands. This is one of the reasons why it is considered a riskier country than Japan, even though it has a lower level of debt to GDP ratio. Many South American countries are closer to the Greek reality than Japan's and as the following section will show, countries' ratings are given accordingly.

1.4.6 Regional credit rating

The capacity of debtors to honour their obligations is measured by credit rating agencies. The most commonly known are Moody's Investors Service, Standard & Poor's (S&P) and Fitch Ratings. They evaluate government and corporate bonds as well as other types of collateralised debt obligations including securities based on residential mortgages. Although they have been operating since the beginning of the last century, their operations and methodology have been severely questioned since the last Global Financial Crisis (GFC) in 2008, when the rating given by these three agencies to securities based on US subprime residential proved to be extremely optimistic (White, 2010).

In any case, the development of ratings implies obtaining a very significant volume of economic and financial information on the issuer and the purpose of the collateralisation. For example bonds issued for new infrastructure projects will be analysed in relation to the project itself, the environment in which it will be developed, and how it will affect the solvency of the issuer in the context of its resources and productivity. The longer the term of the debt to qualify, the greater the volume of information that will need to be assembled and analysed, making it easier to give a rating for a short-term debt rather than a long-term debt.

Ratings usually range from AAA (investment grade) to D (speculative grade). Grades in between these extremes can be qualified further (by a number 1–3 for Moody's or a (+) or (−) sign by S&P and Fitch) to show relative standing to major rating categories.

Prospects for economic growth are usually taken into account for rating countries; for example, if an economy is forecast to grow, then this economic growth is expected to lead to higher tax revenues and reduce unemployment, making it easier to reduce the debt to GDP ratio. The debt interest payments as a percentage of GDP is also taken into consideration; for example, if interests are low, even if the debt to GDP ratio increases, the debt payments can still be manageable. But if interest rates take a sudden upward turn, then countries are faced with the double combination of higher debt to GDP ratio and higher interest payments as a percentage of GDP. Finally, as previously explained, the question of who owns the debt plays an important part.

The political context in a politically charged region such as South America is also important for the rating of countries. For Brazil, during the period of the presidential impeachment and the corruption scandals around the state oil company Petrobras in 2016, the country's long-term debt was downgraded by two rating agencies: S&P downgraded Brazil's long-term rating from BB+ to BB; Moody's from Baa3 to Ba2. In the case of Argentina, the new government took over power in December 2015 and implemented macroeconomic measures that were welcomed by global investors; therefore, by March 2016 all three credit rating agencies (Moody's, S&P and Fitch) had given the country a more positive outlook. Table 1.22 compiles data from all three agencies for the Big Five; Venezuela has been added for comparison as well as the UK and the US.

As Table 1.22 shows, all agencies are in agreement on outlooks for Peru and Venezuela as well as the two advanced economies. Peru's stable rating is mainly due to the political consensus of the macroeconomic policies implemented by the new administration of President Kuczynski, while Venezuela's outlook is based on the risk of default and by a weak institutional government.

The differences in outlooks for the rest of the countries can be classified into two main categories: political risk (Argentina and Brazil), and fiscal risk (Chile and Colombia). For the first two, Moody's more positive stand is based on confidence in the current administrations, while S&P and Fitch's negative rating is based on the uncertainties about the pace of implementation of government's corrective economic plans, given the social pressure for Argentina and in the case of Brazil, the potential government fallout from corruption investigations.

Chile and Colombia are both suffering from a prolonged low economic growth coupled with increases in government debt: a combination that it is weakening

Table 1.22 Credit ratings 2017 and outlook by country and agency

	Moody's	S&P	Fitch	Moody's	S&P	Fitch
Argentina	B3	B	B	positive	stable	stable
Brazil	Ba2	BB	BB	stable	negative	negative
Chile	Aa3	AA-/A-1+	A+	stable	negative	negative
Colombia	Baa2	BBB	BBB	stable	negative	stable
Peru	A3	BBB+/A-2	BBB+	stable	stable	stable
Venezuela	Caa3	CCC	CCC	negative	negative	negative
UK	Aa1	AA/A-1+	AA	negative	negative	negative
US	Aaa	AA+/A-1+	AAA	stable	stable	stable

Source: Moody's, S&P and Fitch April 2017 data compiled by authors

Data: Credit rating for long-term debt by country and agency both governments' balance sheets. In the case of Colombia, S&P is considering lowering the rating if its external balance sheet or its fiscal debt burden do not improve within 2017–18.

In sum, the spectrum of countries presented here show how economics and politics are closely interrelated in the region and how can this affect the sovereign's profiles, international ratings, foreign investment and capital markets. Chapter 6 will look in more detail at the barriers these issues pose for real estate investment.

1.4.7 Bank lending

As previously explained, the need to develop large infrastructure projects has forced many South American governments to issue bonds and promote public bids to bring local and international investment and lessen the burden on the state. In Brazil, funding for long-term projects is provided by the Federal Bank BNDES, while private banks usually provide bridge loans for some part of the project (usually to cover the initial part of the scheme while the large debt is structured with lending agencies). For large projects it is common that bank syndicates are formed, as few single banks will be willing to be exposed to such risks on their own. This is the case, for example, of the Lima Metro loan negotiated and issued in 2014–15 by a banking consortium including Societé Générale and Banco Santander S.A. – Milan Branch, amongst others (Avendaño and Puiggros, 2016). A syndicate of banks operating in large projects has been successfully implemented in developed countries such as Broadgate and Canary Wharf in the UK during the boom years of property development in the mid-1980s.

The creation in 2011 of the infrastructure debentures in Brazil also introduced a new method for long-term finance in the region. Similar to mortgage debentures in the UK, the loan is advanced for the project but provides the lender a recourse against the assets of the company or a parent company (Harvey and Jowsey, 2004). Additionally, Brazil granted tax exemptions to local and foreign investors involved in these infrastructure debentures and this has been very successful in attracting foreign capital (see Chapter 6 for more details on incentives to cross-border investments). According to Simões Ruso and Rodriguez Cruz (2016), Brazil's total requirement for infrastructure funding for long-term projects from 2010 to 2016 was around 70–100 billion of Reales of investment. In Chile the Alto Maipo Hydroelectric Project, a major infrastructure initiative developed near the capital Santiago, was funded with USD 1,217 million involving the IFC, OPIC and IDB. International banks have found in Peru a new financial arena in the region, due to the significant growth of the country's economy in the past ten years. Thanks to this, Peruvian companies have now access to local and international financing without restrictions, which has increased competition amongst lenders (Avendaño and Puiggros, 2016).

Notwithstanding, a recent report by the IMF (2016a) warns that major global banks are leaving the region. According to Fitch, the exodus is mainly due to tightening of international regulations and the need to reduce exposure to regional inherent risks. However, this has opened the door for regional banks (mainly Brazilian and Colombian) that are expanding via acquisitions.

1.4.8 *Mortgages*

In terms of small commercial banks, their activities are mostly confined with investment lending and underwriting for debt instruments. One such example are mortgages, which are loans provided by financial institutions using real estate as collateral. In South American countries, mortgages have a similar format to the ones issued in UK. In general they consist of an agreement in writing between a creditor and a debtor, which is signed by a witness (a minimum of two witnessed signatures is necessary in some countries). The information contained in the documents or deeds records particulars of the collateral, including the value of property, maturity of the loan and interest rates charged. Properties and mortgages must be registered at the relevant land registry office for the jurisdiction where the property is located. In Chile, mortgaged properties must also be registered in the Prohibitions Registry, which contemplates prohibitions to transfer, convey or enter into new contracts with respect to the mortgaged property (Peralta and Yubero, 2016). In Argentina, the registration of a mortgage deed in the Land Registry expires after 20 years and so must be renewed, although given the high interest rates, they seldom go beyond that time (Moreno and Chighizola, 2016). In general terms, no other particular documents are needed unless the mortgage deeds are in a foreign language, which will require translation and certification by a local notary. Brazil is the most advanced of all countries in this regard as it has adopted a new system for deeds written in English (Simões Ruso and Rodriguez Cruz, 2016).

Some new regulations can benefit the real estate market. For example, the tax exemption regime imposed in Chile in 2014 saw a growth of available credit by 4.2% during the year, which was mainly said to come from the RE sector as reported by the Chilean Banks and Financial Institutions Associations (ABIF) (Peralta and Yubero, 2016). The following chapters will provide more details on mortgages and financing products for commercial and residential properties, as well as major infrastructure projects. However, this brief overview has helped to demonstrate the strength of Brazil and Chile's capital markets in comparison to Colombia and Peru, while Argentina has a long way to go in order to control inflation and IR for the development of a RE market with the level of sophistication of the others.

1.5 Supply and demand

In general terms, the law of supply and demand states that when there is high demand for a certain good or service, then its price increases. Conversely, if there is a large supply, the price of that good or service will tend to fall. But unlike other goods, when the demand of properties changes, it takes a long time for the supply to adjust (for example, a higher demand in office space and the time that it takes to build new office buildings). In other words, the supply lags behind the demand. As seen previously in this chapter, factors such as inflation, interest rates and wage levels can have a direct effect on the supply and demand of all kinds of properties and the flow of money for investment. At times of low IR, the market is active, buyers and developers can have access to credit and, provided the market works efficiently, the supply would tend to move towards equilibrium with the demand. But a hike in IR would mean that the demand could immediately halt, leaving perhaps some developers with projects in the pipeline, with the risk of oversupplying the market once the projects are completed.

Additionally, and as seen before, the demand is affected by population growth (more people will need more houses); while changes in characteristics of the demographics can have a more specific impact (a rise in singles due to divorce and/or more elderly people living alone, can increase the demand for smaller residential units). Equally a raise in white collar employment levels can have an impact on office space; while blue collars affect industrial space. Other factors are, for example, the location of the demand. In London, for instance, the economic activity has been moving east of the city, while house building has been west-bound, a trend that can have a negative effect on transport.

These are general definitions and characteristics of the supply and demand in real estate. The next two chapters of the book cover in more detail commercial and residential demand and supply for the countries under study here, while this section intends to cover more general aspects and the particularities of demand and supply at regional level.

1.5.1 Regional characteristics of the demand and supply of real estate in South America

As seen in previous sections, the informal sector of the population plays an important part of the regional dynamics, where some countries have a >45% level of informal economy; while others have between 30–45% (Table 1.11). So these are high percentages of the population that affect the demand of real estate and infrastructure in different ways.

For the residential and infrastructure sector, the trend to upgrade slums rather than relocation and clearance, is a welcome measure as it avoids displacement and potential land conflict. Still, upgrading decisions add a considerable strain to the demand side; for example, the demand of infrastructure for the provision of clean water and sanitation (one of the SDG). Furthermore, the formalisation of entire neighbourhoods constitute an immediate new supply to the housing market that was previously not accounted for within the formal market. Some authors maintain that there is already an informal residential market operating in parallel to the formal (Silva and Mautner, 2016). However, a real estate market as described by these authors, where home owners have to sell their properties because they have had threats from local gang members and are forced to move out of the slum, hardly constitutes the fundamentals of a commercial market where there is a willing seller and a willing buyer. Regardless of the definitions

and whether this constitute a real estate market or not, the crucial and yet unre-
solved question to governments becomes how and to what end the estate should
regulate these markets, particularly in the context of the SDG, which aim at
reducing poverty and deliver sustainable living for all. The convergence of the
two formal and informal markets, including their respective demand and supply,
is something that South American governments must manage through clearly
defined policies.

One of those policies that needs revision is the question of land title, which is
the first barrier informality encounters to be part of a real estate market. A title
provides owners legal rights as well as protection under the laws of a country. The
issue of titling has been extensively researched and contested, mainly since the
publication of Hernando de Soto's theory on reducing poverty by regularisation
of land titles of informal properties and thus offer the possibility of asset collateralisation for entrepreneurial activities. This theory has been refuted by empirical
studies carried out in some South American countries (Gilbert, 2014; Galiani
and Schargrodsky, 2016). In essence, the studies suggest that after titling, there
is no evidence of property owners increasing their access to credit by using their
properties as collateral; on the contrary the vast majority fail to get accepted by
the lenders. However, some of the studies point out other benefits of securing
land titles; for example, increase in the savings capacity of the families that have
a title compared to those who do not have one; the first group also tends to use
better quality of materials for building the houses, and have a better distribution
of resources among families as well as higher levels of education amongst their
children (Galiani and Schargrodsky, 2016).

1.6 Planning regulations

Other factors affecting demand and supply include planning regulations. The
pace of construction of new developments is influenced by flows of capital but
also by the regulatory framework behind land management and the speed and
efficiency in granting planning permissions. The importance of spatial development plans/planning frameworks/long-term visions has been highlighted by the
New Urban Agenda so that infrastructure, housing and growth are aligned (see
Chapter 7).

The premise for good land management is that property development represents a reallocation of resources to increase welfare (Harvey and Jowsey, 2004).
As these authors explain, this is achieved by targeting Pareto-optima situations,
which means that either via the market system within the private sector, or the
cost benefit analysis of the public sector, land is allocated towards its best usage.
The method helps policymakers to make informed decisions, for example, between
allocating a particular site for new houses instead of using it for agricultural activities. The process of making such decisions must always consider what possibilities
the society will be missing by giving preference to one use instead of another.

Land management in South America confronts several inherited problems that
have been deteriorating over the years. First of all is the conflict with indigenous

groups over land tenancy. As seen in the introduction, this relates to issues of poor management of natural resources and land grabbing by large multinational organisations. In this context, the new relation between the region and China, which is a vast consumer of Latin America's natural resources, is of pivotal importance (Borras et al., 2012). The result of the poor management and low enforcement of environmental regulations is widespread land degradation and deforestation. As reported by the World Bank (Berry and Andersen, 2004), 15% of the land area of South America is regarded as badly damaged (compared to 25% in Africa, which is normally considered to be in a severe state). As the report highlights, degraded areas are strongly correlated with rural poverty, given that it is in these areas where most communities depend on land resources for survival.

During the 1960s and 1970s and under the above described conditions, many were forced to move from rural to urban areas in order to find work. In some countries such as Colombia, displacement of rural population was due to over 50 years of conflict, while in other countries, natural disasters coupled with land degradation were the most common push factors. Notwithstanding and as seen earlier in this chapter, the rate of rural to urban migration has been in decline and as shown in Table 1.5, Argentina, Brazil and Chile are over 85% urban, while Colombia and Peru are not far behind at over 75%. As Gilbert indicates (2014) most of the rural to urban migration has stabilised in the region, mainly because communications and transport has improved and rural life is not as isolated as it was in the 1970s.

It can be argued that another factor for the current stagnation of rural to urban migration is due to deteriorating living conditions in cities, particularly considering the poor management strategies for coping with rural migrants in the first place. Since the 1970s the urban sprawl in the region was fuelled by rural to urban migration, but parallel to this was the exodus of the wealthy to the periphery, who intended to combine country living with city work via the connectivity offered by the expansion of motorways. The developers were quick to capture the trend and projected large gated communities in the peripheries of many cities, for exclusive use of medium to high income families. Lack of urban planning meant that municipalities were unprepared for this surge in demand in their jurisdictions, allowing developers of gated communities to take most of the good peripheral land. This scenario not only left the municipal authorities with no land resources for building affordable homes, but also left them with the task of servicing the land (Brain and Sabatini, 2006). The new serviced land dramatically increased the values of peripheral plots, sometimes by over 500% (Smolka, 2013). Land squatting became then the only solution for those who sought a home while simultaneously looking for employment in the new gated communities, mainly in the service and security sector. Informality erupted and spread around enclaves of wealth, usually taking poor quality land that was of no interest for real estate developers. As a result, the urbanisation pattern of South America is nowadays characterised by a sea of informality, with patches of wealth in a fragmented city, where the main connectivity is provided by motorways, mostly controlled by the private sector.

Nowadays, this unplanned growth pattern continues, and is proving difficult to break. This is mainly because the urban periphery of cities is controlled by small cash-struggling municipalities with a large proportion of informal population, that pay little or no tax. In this scenario, when opportunities for gated developments arrive, municipal authorities make many concessions to new developments with the hope that the incoming wealthier population will increase their revenues (Murray, 2016). Another common problem is that developers push to break ground as soon as the land is purchased, arguing that the planning process is lengthy and opaque. This is done with the hope that planning permission will be "solved" at some stage by the local officer (de Duren, 2006; Thuillier, 2005). This situation leaves little room for the municipal authorities to negotiate planning gain, where consent for planning is provided upon the developer meeting some of the infrastructure costs (Harvey and Jowsey, 2004). In this scenario developers rule the land, while the municipality is left paying for the infrastructure.

Notwithstanding, some changes are appearing, such as the territorial law in Colombia that emerged at the end of the 1990s. At the time, local authorities were requested to develop a strategy plan (*Ley de desarrollo Territorial 388, 1997-* still in force) (Alcaldía de Bogota Ley 388, 1997). New municipal funds were also created (*Fondos Municipales de Vivienda de Interés Social*), with the aim of collecting and managing revenues for social housing programmes. While the strategy plan allows municipalities to identify land for social housing, the municipal fund equips them with regulations to pursue development contributions from the private sector. This is done either by the demand of land for social housing from the land owner, or by a development levy paid by developers (Chiape de Villa, 1999; Tachópulos Sierra, 2008). However, there are concerns that the regulation is not fully enforced (Carrión Barrero, 2008) and that municipalities only produce a plan just to comply with the national law.

Still, most authors agree that Brazil is leading the way in development contributions. The Special Social Interest Zones (ZEIS) is a new regulation that requires cities to contribute with a certain percentage of infrastructure and social housing for all new developments projects (Prefeitura de São Paulo Decree 44.667, 2004). Currently this regulation has only been applied in São Paulo but with promising results, given that social housing schemes resulting from these programmes intend to target different sectors of the population, encouraging a more social mix (Budds et al., 2005). The other system used in Brazil is the sale of vouchers for development rights in exchange for infrastructure. In this system, if a developer wants to build above the ratio that the size and location of the plot allows, she or he can negotiate this and obtain vouchers that can be either auctioned in the real estate market or used in other sites. In addition, properties that are unoccupied and are considered suitable for social housing whose owners have a land tax debt, can exchange the property for land tax credit. The legislation means that there are no monetary transactions, which increases transparency in the system.

In Argentina, a new law was implemented in 2017 to allow public private partnerships for affordable housing, whereby the government provides the land and 40% of the costs, while the developer provides 60% of the costs and hands over 20% of the units to the Pro-Cre-Ar Housing Programme run by the State.[16] This is the first experience in capturing developer's contributions in the country, and it has only been implemented in the province of Buenos Aires, which is one of the most progressive governments amongst the country's regions. In other areas of the country, the reality is quite different as most municipalities are struggling with infrastructure (Barreto, 2012). Furthermore, some municipalities are so short of resources that they have started to divert most of the subsidies received by central government for infrastructure to the payment of municipal employees. A similar situation of diversion of subsidies to cover for other shortcomings was seen in Colombia during the time of the Instituto de Crédito Territorial (ICT), the State-run social housing provider that operated until the 1990s reform (Tachópulos Sierra, 2008).

The deviation of funding from housing towards payments for an administrative system that is unable to cope with a large slum population seems to be a recurrent fault throughout the decades in many countries. This lack of resources creates a slow system where planning applications are often delayed, adding pressure on developers who work in the peripheries of cities.

1.7 Chapter conclusions

As seen from the data presented in this chapter, many countries in South America are experiencing economic volatility, inflation and currency devaluations that affect the internal real estate market and increase the risk to foreign investment operations. At global level, the new involvement of China in the regional economics adds another factor to the countries' vulnerability as exporters of commodities.

The persistent low global growth has been sending negative shocks to the region since 2012, which in turn have triggered political unrest and a complete change in governments, with countries like Argentina and Brazil going from extreme left administrations, to more centre-right ones. As seen here, most of these new governments are looking to counteract the negative effects of the commodities downturn, by reducing expenditure and implementing new policies aimed at increasing bank reserves and domestic investment. These are policies welcomed by the international community, but as seen by the rating agencies' section in this chapter, the ability of governments to operate fast and pass and implement reforms is crucial to increase investor's confidence in the region.

The analysis presented here on the depths of the capital markets has clearly left two winners: Brazil and Chile, who are also the only two countries currently hitting their inflation targets. As seen in the cases of Chile and Peru, natural disasters directly affect the countries' GDP and their debt to GDP ratio. Both countries have suffered severe weather events in the recent past, including tsunamis, floods and earthquakes. Chile is again a winner in this sense, being the most resilient and with the best policies for mitigation, shown by the speedy recovery of the countries' outputs immediately after the occurrence of the catastrophes. Peru, who is the most affected by El Niño y La Niña conditions, has yet to implement mitigation strategies to protect the country's food production in order to safeguard national outputs and keep a balanced debt to GDP ratio even in the face of adversity. In this sense, the creation of the Peruvian Fiscal Stabilisation Fund,[17] which seeks sovereign debt reduction, is certainly the right way forward. Finally, interest rates and a stubborn inflation are hindering Argentina's deepening of its capital markets; while Colombia is at the verge of losing is rating due to a combination of bad fiscal policies, increasing debt and lower outputs.

The above study leads to the conclusion that Chile and Brazil can be considered "mature" South American markets, while Argentina, Colombia and Peru can be classified as "emerging" ones. The following chapter will look in more detail at the evolution of commercial and residential real estate markets in the Big Five economies.

Notes

1 Venezuela was originally a full member but has been suspended since 2016. Other countries' memberships are shifting and the block has been expanding. For latest details please refer to the block's website www.mercosur.int
2 www.iom.int/sacm
3 Argentina's manipulation of statistics by the previous administration has been evident, prompting a reform of the Instituto Nacional de Estadística y Censos (INDEC) by the current administration. The country has requested the IMF and other organisations to disregard data prior to 2015 and only in 2016 the international media considered that the country presented reliable accounts http://uk.reuters.com/article/us-argentina-economy-statistics-idUKKCN0ZD1NJ.
4 Eurstats EU28. Available at http://ec.europa.eu/eurostat/statistics-explained/index.php/Unemployment_statistics
5 Financial Times, *Former Contraband Capital Gets Down to Business. Paraguay's Ciudad del Este Changes Its Image as Brazilian Companies Set Up Shop*, 6th April 2017.
6 Measurement of informal economies remains a challenging subject. For a detail study see World Bank Policy Research Studies. Available at http://documents.worldbank.org/curated/en/416741468332060156/pdf/WPS6888.pdf. For Latin America see also from the World Bank. Available at http://documents.worldbank.org/curated/en/326611468163756420/pdf/400080Informal101OFFICIAL0USE0ONLY1.pdf
7 The cost of a Big Mac (and before that of a bottle of Coca-Cola) has been widely used to build comparative indices of cost of living across the world, the most common example is *The Economist's* Big Mac Index (since 1986), originally built to test purchasing power parity (PPP) and test currencies' values, for a detail explanation of the methodology see www.economist.com/content/big-mac-index. The so-called burgernomics has also been used to compare wages of workers doing the same job in different countries, see for example http://s3.amazonaws.com/academia.edu.documents/42381693/ashenfe.pdf?AWSAccessKeyId=AKIAIWOWYYGZ2Y53UL3A&Expires=1492259945&Signature=B2hgE1%2FbFg3HYpFJ9elMc4er8r4%3D&response-content-disposition=inline%3B%20filename%3DCross-Country_Comparisons_of_Wage_Rates.pdf
8 The rental accommodation value provided by Numbeo's website is for a 1 bedroom property with no specifications of square footage, contrary to the South American countries which are all based on the same space.
9 www.eca-international.com/news/november-2016/salary-trends-2016-17-china-release
10 World Bank Gini Index Estimate. Available at http://data.worldbank.org/indicator/SI.POV.GINI, accessed on 5th May 2016.

11   www.latercera.com/noticia/argentina-recauda-us-5-300-millones-blanqueo-capitales/
12   In South America, Colombia's telecommunications company ColTel issued the first
     hybrid in 2015, see ECLAC, 2015.
13   BRF Brazil Foods issued green bonds in 2015, ECLAC (2015).
14   www.latinfinance.com/Article/3646839/Sovereign-Issuer-and-Sovereign-Bond-of-
     the-Year.html#/.WPYqD1LMzBI
15   www.forbes.com/sites/mikepatton/2014/09/29/the-seven-most-indebted-nations/#
     7cbdf5aa69ec
16   www.aevivienda.org.ar/detalle.php?id=585
17   www.swfinstitute.org/swfs/peru-fiscal-stabilization-fund/

Bibliography

Acosta, D. (2016) Free Movement in South America: The Emergence of an Alternative
    Model? *Migration Policy Institute, The Online Journal*. Available at www.migrationpolicy.
    org/article/free-movement-south-america-emergence-alternative-model, accessed on
    31st January 2017.
Arslanalp, S. and Tsuda, T. (2014) Tracking Global Demand for Emerging Market Sover-
    eign Debt. *IMF Working Paper 14/39*. International Monetary Fund, Washington, DC.
Avendaño, J.L. and Puiggros, J.M. (2016) Peru. In *Lending and Secure Finance 2016*. Inter-
    national Comparative Legal Guides Publication. McMillan. Available at www.mcmil
    lan.ca/Files/194273_Lending__Secured_Finance_2016.pdf.
Barreto, M. A. (2012) Cambios y Continuidades en La Política de Vivienda Argentina
    (2003–2007), *Cuadernos de Vivienda y Urbanismo*, 5, 9, pp. 12–30.
Benjamin, N., Beegle, K., Recanatini, F., and Santini, M. (2014) *Informal Economy and
    the World Bank*. Available at http://documents.worldbank.org/curated/en/4167414683
    32060156/pdf/WPS6888.pdf, accessed on 14th April 2017.
Berry, L. and Andersen, T. (2004) *Operationalising GEF's: New GEF OP 15 on Sustain-
    able Land Management Within World Bank Instruments. Latin America & The Caribbean*.
    Washington, DC: World Bank.
Borras, S., Franco, J., Gómez, S., Kay, C., and Spoor, M.L. (2012) Land Grabbing in Latin
    America and the Caribbean. *The Journal of Peasant Studies*, 39, 3–4, pp. 845–872.
Brain, I. and Sabatini, F. (2006) Relación Entre Mercados de Suelo y Política de Vivienda
    Social, *ProUrbana*, 4, pp. 1–13.
Budds, J., Teixeira, P. and SEHAB (2005) Ensuring the Right to the City: Pro-poor Hous-
    ing, Urban Development and Tenure Legalisation in Sao Paulo, Brazil, *Environment and
    Urbanisation*, 17, pp. 89–113.
Carrion Barrero, G. A. (2008) Debilidades del Nivel Regional en el Ordenamineto Ter-
    ritorial Colombiano. Aproximación desde la Normatividad Política Administrativa y
    de Usos de Suelo. *Architecture, City and Environment*, 3, 7, pp. 145–166.
CEPLAN – Centro Nacional de Planeamiento Estratégico. (2016) *Economía informal en Perú:
    Situación Actual y Perspectivas*. Available at http://perureports.com/wp-content/uploads/
    2016/08/Economia-informal-en-Peru-situacion-actual-perspectivas-15-03-2016.pdf,
    accessed on 13th April 2017.
Chiape de Villa, M. L. (1999) *La Política de Vivienda de Interés Social en Colombia en Los
    Noventa*. Santiago de Chile: CEPAL.
Checherita-Westphal, C. and Rother, P. (2012) The Impact of High Government Debt
    on Economic Growth and Its Channels: An Empirical Investigation for the Euro Area.
    European Economic Review, 56, 7, pp. 1392–1405.

de Duren, N. (2006) Planning à la Carte: The Location Patterns of Gated Communities Around Buenos Aires in a Decentralised Planning Context. *International Journal of Urban and Regional Research*, 30, 2, pp. 308–327.

ECLAC – Economic Commission for Latin American and the Caribbean. (2015a) *Statistical YearBook for Latin America and the Caribbean*. Washington, DC: United Nations.

ECLAC – Economic Commission for Latin American and the Caribbean. (2015b) *Social Panorama of Latin America*. Washington, DC: United Nations.

ECLAC – Economic Commission for Latin American and the Caribbean-Washington Office. (2015) *Capital Flows to Latin American and the Caribbean. 2015 Overview*. Washington, DC: ECLAC.

ECLAC – Economic Commission for Latin American and the Caribbean-Washington Office. (2016) *Capital Flows to Latin American and the Caribbean. Q1 2016*. Washington, DC: ECLAC.

Ferreira, F.H.G., Lopez-Calva, L.F., Lugo, M.A., Messina, J., Rigolini, J., and Vakis, R. (2013) *Economic Mobility and the Rise of the Latin American Middle Class*. Washington, DC: World Bank.

Galiani, S. and Schargrodsky, E. (2016) Urban Land Titling: Lessons from a Natural Experiment. In Birch, E., Chataraj, S., and Watcher, S. (eds.), *Slums: How Informal Real Estate Markets Work*. Philadelphia: University of Pennsylvania Press, pp. 83–93.

Gilbert, A. (2014) The Urban Revolution in Gwynne, R. and Cristobal, K. Latin America Transformed. Globalisation and Modernity. Routledge, Oxon and New York. Second Edition, pp. 134–162.

Gómez Sabaini, J.C. and Jiménez, J. P. (2012) *Tax Structure and Tax Evasion in Latin America*. CEPAL. Available at http://repositorio.cepal.org/bitstream/handle/11362/5350/1/S1200023_en.pdf, accessed on 17th April 2017.

Harvey, J. and Jowsey, E. (2004) *Urban Land Economics, Sixth Edition*. London: Palgrave.

ILO – International Labour Organisation. (2016) *Labour Overview, Latin America and the Caribbean*. Lima: ILO/Regional Office for Latin America and the Caribbean.

IMF – International Monetary Fund. (2009) Data Spotlight: Latin America's Debt. *Finance and Development*, 46, 1. Available at www.imf.org/external/pubs/ft/fandd/2009/03/dataspot.htm, accessed on 20th May 2016.

IMF – International Monetary Fund. (2014) Is It Time for an Infrastructure Push? The MacroEconomic Effects of Public Investment. *Word Economic Outlook*. Washington: IMF, October, pp. 75–114.

IMF – International Monetary Fund. (2016a) *Financial Integration in Latin America*. Available at www.imf.org/external/np/pp/eng/2016/030416.pdf, accessed on March 2017.

IMF – International Monetary Fund. (2016b) *Regional Economic Outlook: Western Hemisphere. Managing Transitions and Risks*. World Economic and Financial Surveys.

IOM. (2014) Estudio sobre Experiencias en la Implementación del Acuerdo de Residencia del MERCOSUR y Asociados. Documento de Referencia. Presented at the XIV *Meeting of the South American Conference on Migration in Lima*, October 16–17, 2014.

Moreno, J. and Chighizola, D. (2016) *Argentina in Lending and Secure Finance 2016*. International Comparative Legal Guides Publication.

Murray, C. (2016) Real Estate and Social Inequality in Latin America: Approaches in Argentina, Brazil, Chile and Colombia. In Abdulai, R., Obeng-Odoom, F., Ochieng, E., and Maliene, V. (eds.), *Real Estate, Construction and Economic Development in Emerging Market Economies*. London: Routledge, pp. 256–276.

Murray, C. and Clapham D. (2015) Housing Policies in Latin America: Approaches in Argentina, Brazil, Chile and Colombia. *International Journal of Housing Policy*, 15, 3 pp. 347–364.

OECD – Organization for Economic Cooperation and Development and OAS-Organization of American States. (2015) *International Migration in the Americas: Third Report of the Continuous Reporting System on International Migration in the Americas (SICREMI)*. Washington, DC: Organization of American States.

Peralta, D. and Yubero, E. (2016) Chile. In *Lending and Secure Finance 2016*. International Comparative Legal Guides Publication.

Pickett, K. and Wilkinson, R. (2009) *The Spirit Level: Why Equality is Better for Everyone*. London: Penguin.

Pokorny, B., Scholz, I., and de Jong, W. (2013) REDD+ For the Poor or the Poor for REDD+? About the Limitations of Environmental Policies in the Amazon and the Potential of Achieving Environmental Goals Through Pro-poor Policies. *Ecology and Society*, 18, 2, p. 3. Available at www.ecologyandsociety.org/vol18/iss2/art3/?utm_source= REDD%2B+Digest+-+2+August+2013+&utm_campaign=REDD+digest+05-02-13&utm_medium=email.

Powell, A. (2016) Time to Act. *Latin America and the Caribbean Facing Strong Challenges: 2016 Latin American and Caribbean Macroeconomic Report*. InterAmerican Development Bank.

Rule, G. (2015) *Understanding the Central Bank Balance Sheets*. Centre for Central Banking Studies, Handbook 32, Bank of England.

Silva, P. and Mautner, Y. (2016) Tenure Regularisation Programmes in Favelas in Brazil. In Birch, E., Chataraj, S., and Watcher, S. (eds.), *Slums: How Informal Real Estate Markets Work*. Philadelphia: University of Pennsylvania Press, pp. 83–93.

Simões Ruso, R. and Rodriguez Cruz, L.B. (2016) Brazil. In *Lending and Secure Finance 2016*. International Comparative Legal Guides Publication.

Smolka, M.O. (2013) Implementing land value capture in Latin America (Policy Focus Report). Policies and tools for urban development. Cambridge, MA: Lincoln Institute of Land Policy.

Stieglitz, J. (2013) *The Price of Inequality*. London: Penguin.

Tachópulos Sierra, D. (2008) El Sitema Nacional de Vivienda de Interés Social (1990–2007). In Ceballos Ramos, O. (ed.), *Vivienda Social en Colombia: Una Mirada Desde su Legislación 1918–2005*. Bogota: Editorial Pontificia Universidad Javeriana, pp. 181–238.

Thuillier, G. (2005) Gated Communities in the Metropolitan Area of Buenos Aires, Argentina: A Challenge for Town Planning. *Housing Studies*, 20, 2, pp. 255–271.

Vuletin, G. (2008) Measuring the Informal Economy in Latin America and the Caribbean. *IMF Working Papers*, 8, 102, pp. 1–31.

Weidner, J. and Williams, J. (2011) What Is the New Normal Unemployment Rate? *FRBSF Economic Letter*. Available at www.frbsf.org/economic-research/files/el2011-05.pdf, accessed on 13th April 2017.

White, J. (2010) Markets. The Credit Rating Agencies. *Journal of Economic Perspectives*, 24, 2, pp. 211–226.

World Bank. (2013) Shifting the Gears to Accelerate Shared Prosperity in Latin America and the Caribbean. *International Bank for Reconstruction and Development*. Available at http://documents.worldbank.org/curated/en/132921468012674913/pdf/785070WP0PL B0S00Box377346B00PUBLIC0.pdf, accessed on 17th April 2017.

2 Commercial real estate in South America

2.0 Introduction

From a developer's perspective investing in real estate means losses in liquidity, as financial resources are transformed into land and construction materials that are required for the building process. This original investment can eventually result in a valuable asset and the developer is able to recover liquidity by means of selling or letting the real estate asset. Development activities require local knowledge, from knowledge of best opportunities for land acquisition, to experience of the planning process, construction regulations and standards. Additionally, a good track record in successfully delivering projects can increase a developer's reputation and his/her ability to attract investment and secure leverage from the banking sector.

The information used for decision making usually comes from data that is systematically collected; for instance from reliable real estate agencies such as those operating in advanced economies, CBRE, JLL etc. In comparison, data collection by these agencies in South America is a relatively new activity, which is carried out by separate organisations who have their own methodology and standards. When possible, data collected for this chapter comes from the same source to allow for a comparative analysis between the countries under study here. More details on the opacity of the South American markets due to information asymmetries is presented in Chapter 6, along with an analysis of how this issue can affect investors' and developers decisions as well as the competitivity of the region as a whole in the global scene. In this chapter, the information presented will be used to explain how commercial real estate markets operate in South America; therefore, the quality of the data will not affect the purpose of this chapter as it is presented for illustration purposes and not for market analysis.

Developing countries, such as the South American ones studied here, are characterised by volatile economies and fluctuating consumer spending. Therefore, investment in commercial real estate, which is usually long term, can add a risk factor to the operation. Indeed as explained in the previous chapter, the state of the local economy, the level of employment, interest rates and inflation can affect consumer's spending. This in turn has an impact on businesses and on the demand for commercial space. In a bullish economy, and supposing perfect market conditions exist, unemployment is low, credit is available, and consumer spending is high. This scenario can drive businesses to expand and therefore increase the need for more business space. Additionally, developers' confidence on return on investment is stronger and new projects get underway. Conversely, a bearish economy will have less consumer spending and less demand for commercial space and new projects. Considering that development of new space can take a couple of years to complete, from decision making and land purchase, to planning permission and construction, the volatile South American conditions where the two scenarios of bullish and bearish markets alternate quite frequently, can add a considerable risk to investors. Additionally, in the current context of financial globalisation, any new development proposal must take into consideration alternative lower risk offers available to investors in other markets; for example, the US one, which has a strong regional influence over South American investors (see Chapter 6). Any expectation for an increase in treasury bonds returns in the US will drive investment away from developing countries as the former will be perceived as a lower risk alternative.

In addition to this, and in particular reference to the commodities-oriented economies of South America, any changes in the global demand for raw materials can affect their economic performance and volatility. Furthermore, as shown in the previous chapter, the increasing regional exposure to consumer giants like China (Table 1.1) can also affect the region and compromise local real estate development. Finally, the political scene can affect investors' decisions, as rating agencies monitor political environments in order to assess availability of debtors to meet obligations. As seen in Chapter 1, the low rating of countries such as Venezuela's CCC rating given by all three big agencies, is mainly due to the low credibility placed on the current government, as well as the low productivity of the country, which is affecting its debt to GDP ratio and increasing the likelihood of a default. It is important therefore to highlight that for real estate investors, conscious of the long-term pay-back associated to real estate development, the above considerations, as well as market transparency are essential when making investment decisions.

2.1 Commercial real estate

In this volume, commercial real estate means "non-residential" and within this sector, this chapter will focus on office and retail space. It does not include industrial space, which is used for the manufacturing of goods and does not include hotels and other facilities that are dedicated to recreational and leisure activities. In the Latin American context as a whole, industrial real estate is better represented in countries like Mexico and Panama that attract more FDI than the rest of the region for this activity; while the Caribbean tends to attract more FDI for real estate dedicated to the hotel sector (see Expert box 6.1). For reasons of space, this book only studies South American countries and selects the real estate commercial activities that better represent them, i.e. office and retail.

The position that real estate businesses take in relation to building assets can vary depending on whether they are owner occupiers or tenants. In the latter scenario, an investor owns the property and uses a lease or contract agreement to let the premises to a potential tenant. The contractual agreements in South American countries are usually for a minimum term of two to three years and can be renovated for the same period of time or can be annually revised as it is the case with small retails space in commercial locations across many countries. Leases can also have risk-sharing characteristics (for example, the percentage rent model in the case of shopping centres that will be explained later in this chapter). The advantage many companies seek in renting is to increase their capital for business operations, as the capital that is locked when owning a building is instead diverted towards a company's business activity with the hope that this can enhance their internal operations and overall performance. In addition to this, development of new commercial space requires specialised knowledge and expertise, which are usually not common across all business sectors. Therefore, the most usual scenario is for businesses to rent space rather than develop or be proprietors. The increasing number in sales-and-leaseback operations (i.e. a building is sold to an investor who then leases the premises out to businesses) in South America can testify to the increased preference to rent. In Brazil, for example, the sale and leaseback operations that range from USD 10 million to 200 million in size, have increased by 10% year on year during the period from 2012 to 2015.[1] This represents a considerable amount of new commercial space (office and retail) added to the Brazilian rental market.

The process of capital demobilisation described above where businesses become tenants, allows for the management of the property to be transferred to a landlord. This has another advantage to businesses, as they are able to reduce the need for skilled personnel required for building management. Additionally, and in the case of offices, the task of developing internal spaces completely fitted to the tenant's specifications can be transferred to specialised firms by means of a built-to-suit lease, relieving companies from having to hire design experts to fulfil this activity. The separation between development and management of commercial space has allowed for professional specialisation by distinguishing companies that are focused on commercial real estate development from those dedicated to asset management.

This change of paradigm has already taken place in most of South America's real estate markets, and as a result, specialised international companies have landed and settled in the region, including CB Richard Ellis, Cushman & Wakefield, Jones Lang-La Salle, Hines, and Colliers. Their presence is certainly a contributing factor to the increased sophistication of the market. They also provide valuable market data for international investors, which is gradually improving market transparency.

Because of the distinctiveness of the development process and the lease structures that characterise office and retail space, this chapter will focus on each segment separately, covering first the development process of office space, including location and building standards, followed by a similar study on retail space and in particular on shopping centres. The concluding remarks will summarise the state of both markets in South America.

2.2 Office space in South America

In order to provide the reader with a good understanding of the office market in the region, this section of the chapter is further subdivided into three subsections. The first one (2.2.1) presents the geographical location of office hubs, and considers aspects at neighbourhood and city level. The second (2.2.2), relates to the buildings and its internal features in order to understand tenants' demands and design standards that are typical in South America. A final section (2.2.3), presents an analysis of the region's office market using data from Cushman & Wakefield and JLL's market reports.

2.2.1 Geographical location of offices

According to data collected from the Population Division of the United Nations, there are three megacities (with population of 10 million or more) in South America: Buenos Aires, Rio de Janeiro and São Paulo (Table 2.1). There are also three cities that are over 5 million: Bogota, Lima and Santiago de Chile, which are also capital cities. The second largest urban agglomerations that are not megacities in the Big Five economies include: Medellín in Colombia (near the 4 million mark) and Córdoba in Argentina (over 1.5 million). The last two secondary cities are Valparaíso in Chile and Arequipa in Peru (both under 1 million inhabitants). In Latin America as a whole, there are 198 large cities (defined as having a population of 200,000 inhabitants or more), which together contribute to over 60% of the regional GDP (McKinsey Global Institute, 2011). The cities in Table 2.1 are amongst those 198 Latin American cities and they certainly represent the centres where economic activity is the highest within the South American office sector.

Table 2.1 Urban population in largest cities

City	Population	Category
São Paulo (Brazil)	21,066	megacity
Rio de Janeiro (Brazil)	12,902	megacity
Buenos Aires (Argentina)	15,180	megacity
Córdoba (Argentina)	1,511	secondary city
Santiago de Chile (Chile)	6,507	capital city
Valparaíso (Chile)	907	secondary city
Bogota (Colombia)	9,765	capital city
Medellín (Colombia)	3,911	secondary city
Lima (Peru)	9,897	capital city
Arequipa (Peru)	850	secondary city

Source: United Nations, Department of Economic and Social Affairs, Population Division (2014). World Urbanization Prospects: The 2014 Revision. Statistical annex to the 2016 World Cities Report

Data: Population total (million)

According to some authors, the demand for office space in cities expressed in terms of absorption rate and the rental prices of offices, is strongly correlated to economic activity and GDP (for empirical evidence, see, for example, in developed countries Rosen, 1984; Wurtzebach et al., 1991; Dobson and Goddard, 1992; D'arcy et al., 1997; Liow, 2000; Brooks and Tsolacos, 2001; Bennet, 2005; Geltner and Miller, 2006; in developing economies, Hui and Yu, 2006, for Asia and Rocha Lima Jr. et al., 2016, for South America). The location of major Central Business Districts (CBDs) in the Big Five are located in capital cities, which are also the economic drivers of the countries, such as La City Porteña in Buenos Aires; Sanhattan in Santiago de Chile; the Centro Internacional in Bogota, or the San Isidro District in Lima. In Brazil the new capital, Brasilia, was developed in the 1960s to host political and administrative offices, but financial and business activities were kept in the previous capital, Rio de Janeiro, and specially in São Paulo, the great South American business hub. During the 1970s, São Paulo developed the Avenida Paulista, an avenue that crosses the city on a north-west to south-west direction at one of the highest points in the city's undulating topography. It was roughly at the same time that Buenos Aires developed the Complejo Catalinas, which began in 1969 in an area known as Bajo Porteño. The aim of both cities, São Paulo and Buenos Aires, was to provide a response to the demand for a new type of office space generated by the dawn of the computer age (see next section). The constant trend of cities playing catch up with changes in office design brought by new technologies, alongside increased traffic congestion in city centres, meant that these CBDs, built in the 1970s, were followed by others built in the 80s and 90s, such as Itaim Bibi and Marginal Pinheiros in São Paulo.

The next large secondary city where there is an important commercial real estate activity is Medellín, which has the largest population amongst the non-megacities described in Table 2.1. At country level, Medellín has nearly a third of the population of the capital and therefore presents a more challenging competition to Bogota in terms of office space on offer in Colombia (see Graphic 2.1). This is mainly due to the complete transformation that the city underwent since the radical work implemented by a string of consecutive mayors with a shared vision on how to overcome the stigma left by the infamous Cartel de Medellín that operated in the city during the 1980s and 90s. To combat this, Medellín implemented the so-called social urbanism, which aimed at integrating the poorest neighbourhoods located in the periphery with the city centre. The famous Metrocable (see Chapter 5) is just one of the many examples of the urban revolution that Medellín's mayors put into motion after the turn of the century.

The demand for office space in Medellín came via the powerful Grupo Empresarial Antioqueño (GEA), an association of over 100 companies that are based in the Department of Antioquia, of which Medellín is the capital. The GEA reportedly generates 7% of Colombia's national GDP[2] and with the support of innovation centres such as RutaN,[3] has helped to catapult the city as a destination for international companies willing to expand in Colombia. Geographically, businesses concentrate to the east of the city, in an area known as *El Poblado*, which is also attracting commercial activity in the form of restaurants and bars around the Calle 10.

The presence of office buildings can indeed instigate a surge in surrounding amenities (food quarters, shopping areas, recreational open spaces, etc.) as it happened with El Poblado. The amenities can enhance the overall quality of the neighbourhood and attract more office developments as an increasingly sophisticated and international workforce is valuing quality of life as well as career moves, making the most of the locational choices offered by global businesses and economies (Clark et al., 2002; Florida, 2005). The surge of amenities around office locations can be a spontaneous process driven by the entrepreneurialism of the locals who are keen to cash in on the opportunities brought about by office employment activity, particularly during lunch break and after work hours. This urban transformation usually takes the shape of incremental changes that follow fashionable trends. For instance, the demand for a certain location can be the result of artists as well as media and advertising industries moving into a derelict area seeking lower rental values. Shoreditch, in London, is certainly a well-known example of this urban transformation process. Alternatively, the development of urban amenities can be actively pursued by planning officials via a regeneration project. For example, a local authority can instigate the development of a new CBD to reduce congestion in the old city centre, or it can be a business case put forward by multisector groups as part of a wider regeneration plan of an old industrial area. Examples of this in Europe are La Defénce in Paris and Canary Wharf in London.

In South America, Buenos Aires presents a case of each type of development: entire regeneration projects, incremental urban changes to the creation of a new business district; starting with the development and regeneration of Puerto Madero, the bottom up transformational urbanism of Palermo Hollywood, and the government led creation of a business district in Parque Patricios. Puerto Madero saw the regeneration of the degraded old port, which housed disused warehouses and underdeveloped land. This was transformed into a modern business and residential district, using federal and city resources. A joint-stock company was created with the objective of urbanising the area, with the transference of the totality of the space to this company after completion of project in a process that was regulated by the urban development authority. An example of gradual and incremental transformation can be seen in the area now known as Palermo Hollywood where businesses slowly moved into the area during the 1990s lured by lower rents and easy access to the City Porteña. Finally, the initiative under the Kirchner administration (2003–15) to create a technology hub in Parque Patricios, initiated a new policy now followed by the current Government, to develop specialised districts in specific locations of the city including a design district, an art district and a sport district.

The development of special districts is also seen in other cities in South America. For example, in São Paulo, Brazil, the local authority can earmark an area through a development tool known as "Operação Urbana" – Urban Operation.

Figure 2.1 Puerto Madero
© Claudia Murray

The innovative aspect of this urban operation is that it allows the local authority to capture developer's contributions and re-invest them in the same area.
The first step of the urban operation is that the local authority defines a specific
interest zone with the aim of transforming the area through a public-private-
partnership. The planning authorities then design a master plan with regulated
building densities and heights. If the developer wants to build higher than what
it is stipulated in the new plan, they can opt to purchase the airspace rights above
the proposed buildings from the local authority but only up to a certain agreed
limit. This purchase is done by means of auctions in the open market. The money
captured by the selling of the rights is exclusively directed to benefit the area
by adding infrastructure or other improvements to the area as predetermined in
the master plan. The title of investment acquired in auctions is called CEPAC –
Certificate of Additional Building Potential – and is issued within the limit of
additional construction for the area. São Paulo was a pioneer in the use of such
an instrument and the Urban Operation denominated *Operação Urbana Con-
sorciada Água Espraiada* was one of the most significant.[4] The total area of the
project was 3.75 million square metres, including 30% for residential use and
70% for commercial and other uses – for example community infrastructure – as
prescribed by the plan. Several area enhancement elements were implemented

such as the Octávio Frias de Oliveira bridge, now an important landmark in São Paulo but also a crucial link improving connectivity on both sides of the river. The urban operation also worked with a local community of slum dwellers, who were temporarily removed to allow for the building of new housing. The housing complex known as Jardim Edite also includes a hospital and a kindergarten for the community.

In Santiago, the new business district and financial core of the city, the area now known as Sanhattan and the Costanera Centre, was developed to ease congestion in the old commercial part of the city. The selected site for the regeneration project was the former site of Chile's national brewery (CCU for its Spanish acronym). The area was subdivided into plots and gradually bought by private developers who first saw the potential of the location in terms of connectivity and easy car access, a complete change from the narrow roads of the old urban core. The first buildings appeared at the end of the 1990s and were also a design revolution as they introduced large glass curtain-walls following the style of office buildings in Manhattan, NYC. The nickname given to the area by the locals, Sanhattan, appeared as a combination of Santiago and Manhattan in reference to the stylistic similarities. The Costanera Centre is a more recent development which now has both the tallest building in South America (Gran Torre Santiago)

Figure 2.2 Frias de Olivera Bridge
© Claudia Murray

Figure 2.3 Jardim Edite
© Claudia Murray

as well as the largest shopping mall. Paradoxically, the project that was originally developed in order to offer solutions to the congested old centre, has now generated severe traffic problems in the Providencia neighbourhood where it is located.

As also mentioned in the case of Palermo Hollywood in Buenos Aires, regeneration and clustering of business activities can happen incrementally and spontaneously. Usually this gradual change is followed by developers' interventions who are quick to step-in in the area after spotting particular urban trends and small businesses' preferences. Many researchers have highlighted this increase in the relocation of office in residential neighbourhoods in many cities around the

Figure 2.4 Torre Santiago
© Claudia Murray

world (Archer, 1981) Cases in South America include San Isidro and Miraflores in Lima and the aforementioned El Poblado in Medellín. The advantage of this gradual change is that it generates a balanced mix of business and residential activities, promoting the use of valuable urban space for a longer period of time during the day, and not just during the 9–5 hours of the working week. This can increase the usage of core urban areas, which are usually well serviced by public transport as well as discourage criminal activity due to higher human traffic at all hours and specially weekends. The mix of residential and work activities will therefore be beneficial for South American cities, which suffer from both defective transport links and high levels of criminality (see Table 2.2 in the next section).

The disadvantage is that in most cases, such as that of Providencia in Santiago, the local authority is slow to react to the changes and misses the opportunity to capture the value uplift of the land in order to build resources to invest in infrastructure projects that the new surge in activity demands in transformational urbanism. The lack of control in the development process in Chile is a recurrent problem that has been associated to the control exerted by certain individuals with vested interests (including political figures) in the decision making process of the city's urbanisation process (Zunino, 2006).

2.2.2 Office buildings' specifications

There are different needs in terms of building specifications that might be required by tenants, which is related to the type of industry they are involved. For example, law firms might have different requirements than media firms. Additionally, today's modern technologies are fast moving and this can affect the speed of obsolescence of an office building. For instance, when the IT industry revolutionised businesses' operations, many office spaces built before the 1980s were not able to adapt to the demand for the larger floor-to-floor minimum height required to install all the added electrical connections needed for computers. Furthermore, as most of South America's office space is located in the tropics, complex air-conditioning systems are usually added to the list of end-users' demands, and this also requires additional space for fitting conduits (see Graphic 2.1).

Other important components to office specification that are relevant to the South American market are security and parking requirements. The former is mainly due to the high levels of insecurity and criminality prevalent in the region (Table 2.2), while the latter relates to the lack of reliable public transport affecting most major cities (see Chapter 5). The combination of the two, the need for parking and security, has led to a trend in the design of office buildings in isolation from the rest of the neighbourhood and with very little permeability as buildings tend to be surrounded by high walls or security gates.

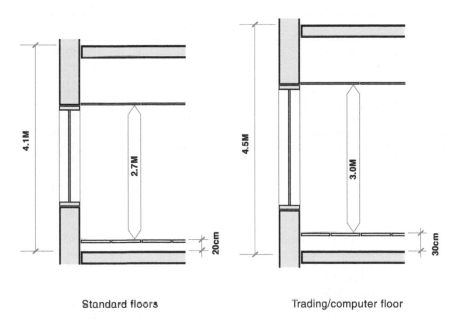

Standard floors   Trading/computer floor

Graphic 2.1 Floor to ceiling height

© Ween Architects

Current efforts in increasing transport connectivity are helping to reverse the trend in car parking requirement, but security issues remain. Table 2.2 presents the ranking of cities according to crime rates collected by Numbeo. The agency conducts regular surveys in major cities around the world to rate the level of insecurity amongst the population. The data is not based on actual information of crime rates, but gives an indication of the perceived insecurity amongst the local population. This is important as usually perceived fear of crime tends to be higher than the actual levels of criminality in a given area. Therefore, crime perception can affect people's attitudes and behaviour more than real statistics (Skogan, 1986). Perception of crime will be therefore more likely to influence the building specifications from potential office tenants rather than the actual data on crime.

As a result of this large range of office specifications, from ceiling height to security and parking, combined with the aforementioned appearance of international real estate agencies and consultants in South America, new office standards and classifications were gradually introduced in the region since the late 1990s.

There is a considerable debate over the classification of the buildings and design standards as these can be based on subjective judgements but can affect the sale price of assets (Fuerst et al., 2011). In the case of the South American office

Table 2.2 Ranking of cities by crime perception

Rank	City	Crime index	Safety index
1	Caracas, Venezuela	86.61	13.39
2	Fortaleza, Brazil	83.9	16.1
3	Recife, Brazil	78	22
4	Rio De Janeiro, Brazil	77.87	22.13
5	Porto Alegre, Brazil	77	23
6	São Paulo, Brazil	72.19	27.81
7	Rosario, Argentina	69.14	30.86
8	Belo Horizonte, Brazil	68.96	31.04
9	Lima, Peru	68.27	31.73
10	Campinas, Brazil	66.8	33.2
11	Curitiba, Brazil	65.16	34.84
12	Brasilia, Brazil	64.52	35.48
13	Bogota, Colombia	63.64	36.36
14	Goiania, Brazil	62.12	37.88
15	Buenos Aires, Argentina	61.11	38.89
16	Montevideo, Uruguay	55.73	44.27
17	Florianopolis, Brazil	55.36	44.64
18	Quito, Ecuador	54.96	45.04
19	Santiago, Chile	52.55	47.45
20	Medellín, Colombia	50.73	49.27

Source: Numbeo Crime Index 2017

Table 2.3 Co-working space by country

Country	Number of addresses	Number of cities
Argentina	11	7
Bolivia	1	1
Brazil	23	17
Colombia	8	3
Chile	5	3
Ecuador	1	1
Peru	4	1
Uruguay	1	1

Source: Coworking 2017

Data: www.wikicoworking.org

market the classification is important as the region has generally been known for its lack of Class A office space if compared with similar space offered in the developed world. This is mainly due to the tendency towards owner-occupation that is prevalent in the region. Indeed owner-occupiers have less resources to invest in the upkeep of their offices and can contribute to the early obsolescence of the building due to lack of investment for refurbishment projects. Notwithstanding and as stated in the introduction of this chapter, the increase of sale and lease-back operations seems to indicate that the region is gradually converging with other world cities, constantly adding new developments under single ownership that allows for better maintenance standard of assets.

South American countries are also catching up with the trend of co-working in office space. This is defined as the sharing of a working environment by multiple tenants. The space provided has all the facilities of an office but has the advantage of making the rents more affordable. As a result, these spaces cater for the self-employed and small entrepreneurial businesses. In total there are 54 locations across South America that can offer co-working spaces (Table 2.3).

2.2.3 The South American office markets

The income stream received by commercial property investors is regulated by the lease contract and the rights agreed by the two parties: owners and occupiers (Baum, 2009). As this author explains, the characteristics of the lease – for example, length of contract, break clauses, rent reviews, as well as repairing liabilities and use and assignment restrictions – can affect the performance of the interest in the building. It is therefore important to consider the characteristics of the commercial leases across the countries under study in this chapter.

According to JLL (2016) commercial leasing practices are remarkably similar across many South American countries. For example, lease length can be between three and five years, although some countries such as Brazil, Colombia and Uruguay have a typical term of five years, while Venezuela has a more variable term, possibly reflecting the unstable economic conditions and the unwillingness of tenants to sign for longer terms. The rents in most countries are fixed for a year and indexed adjusted upon renewal, contrary to the UK, where the current system has fixed rents for five years and are discretionally adjusted (although they cannot be adjusted for a lower value of the first contract; this is called an upwards-only review). In South America's more volatile economies, with higher levels of inflation than the UK, rents are adjusted annually, and have no upward-only review clauses in the contracts.

Currency is an additional factor that is not considered in either the UK or the US. In Brazil, Chile, Colombia, Peru and Venezuela, rental prices are stipulated in the local currency, while Argentina tends to use US dollars, reflecting the instability surrounding the local currency. Repairing liabilities are the responsibility of the landlord, while service charge is paid in advance by the tenants to the management company. Most countries accept the inclusion of break clauses in the contracts with the exception of Colombia and Ecuador, where early termination of the lease is not possible and contracts must be fulfilled for the whole term agreed. A distinctive characteristic of South American leases is the prevalence of the provision of parking spaces. Indeed as previously explained, most office workers still rely on their cars as their main mode of transport for the daily commute. As a result, contractual rental agreements tend to offer parking spaces in addition to the office space. These are usually calculated as one space for every 35 to 100 sq. m of rented office space.

Apart from the lease structure, the strength of an office rental market is measured by key market indicators, which include: a) the inventory, b) vacancy rate, c) absorption and net absorption and d) prices. As explained previously in this chapter, office data available in each of the most important cities in South America is not homogeneous. For example, some countries' longitudinal information is patchy or is very recent (Argentina and Brazil have the longest series going back to the late 1970s while Chile, Peru and Colombia are more recent). At the city level, São Paulo, Buenos Aires and Rio de Janeiro have the longest longitudinal data, and they are also amongst the largest office markets in South America. Their level of data availability are followed by Bogotá, Santiago and, more recently, Lima.

Using data available for the most sophisticated office space available in each location, the rest of this section will present: i) the inventory or total stock; ii) the vacancy rates; and iii) the net absorption in order to explain how political and economic conditions are closely interlinked with the office rental market in the Big Five.

i) The inventory or total stock

The total stock of a rental market shows the square metres of office floor in a city, no matter if occupied, available or being refurbished. As explained in Chapter 1, the inventory can be considered a lagging indicator due to the fact that it takes several years to build new office buildings and increase the supply of space. The

inventory therefore can be better understood in the context of other market indicators, which are more dynamic; for instance, vacancy rate, which shows take up
of office space by tenants. Many commercial real estate agencies and consultancy
firms operating in South America produce their own market indicators. There
are therefore some differences when reporting the inventory, so in order to allow
for cross country comparison, data for this chapter has been collected from one
real estate consultant (Cushman & Wakefield).

Graphic 2.2 shows the size of the different office markets for qualified stock
across South America for 2015 as well as the space expected to be delivered
between 2016 and 2017. Data is for rental area only and as reported by JLL for
the year 2015.

As seen in the graphic, São Paulo is the biggest market with nearly 4.5 million square metres of office space available for rent, with an expectation that
the city will surpass the 5 million mark at the end of 2017. The second largest
rental market for office space is Santiago de Chile, with just over 3 million square
metres of office space considering currently available space and expected new
supply. The capital city of Colombia, Bogotá, is expected to reach the 3 million
mark by the end of 2017. Rio de Janeiro, the second largest market in Brazil, has
a current rentable stock of just under the 2 million mark but new supply will put
this city over this threshold. Not surprising, the rest of the largest markets are
located in capital cities across the region (Lima, Quito, Buenos Aires, Caracas,
and Montevideo), with each having a total rentable stock of less than 1.5 million square metres of office space. Amongst all of them, the highest growth in
supply is expected in the capital of Peru, Lima. The rest of secondary cities in
graphic 2.2 are distributed in the Andean countries, two in Colombia (Medellín and Cali) and one in Ecuador (Guayaquil). The presence of these cities in

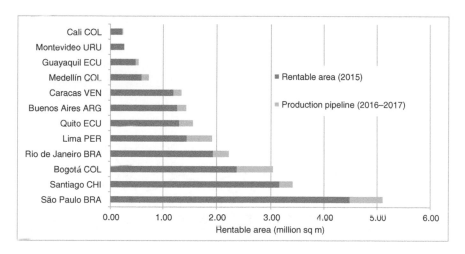

Graphic 2.2 Rentable areas and new supply

the graphic reflects the importance of the local economic activity. In the case of Guayaquil, this is one of the most prominent ports in South America which helps to serve the Pacific Ocean trade. The city is also the commercial heart of the country, unlike the capital Quito, which has more governmental functions. As explained in the previous section of this chapter, Medellín is the second most important economic centre of Colombia, being Cali the third one and according to the municipal authorities of the city, contributing with nearly 6% towards Colombia's GDP.[5]

ii) Vacancy rates

Vacancy rate is one of the most important indicators of office market conditions, as it points out the segment of the office floor space that is vacant at a given time. Depending on market size and all things being equal, the lower the vacancy rate, the higher is the demand for space. Vacancy rate is always referred to a specific point in time, and is its longitudinal behaviour along with the other indicators (such as absorption and net absorption) that help to build a forecast for a particular office market.

Graphic 2.3 shows a comparative chart of vacancy rates as percentages of total stock for class A office space in South America's largest markets. Historical data has been collected from Cushman & Wakefield for most cities with the exception of Brazil where data has been collected from the Buildings database. From the years 2000 to 2005, only data for Buenos Aires is available and it clearly shows how the country's default in 2001 triggered a peak of vacancy rates that reached

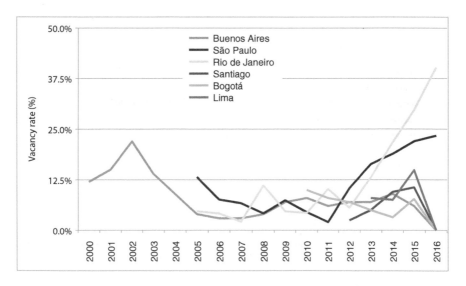

Graphic 2.3 Vacancy rates

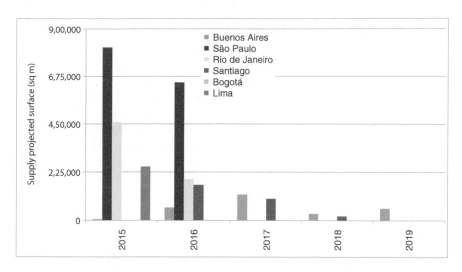

Graphic 2.3 (Continued)

22% in 2002. There was a slow recovery from 2005 to 2008 during the years of president Néstor Kirchner (2003–07), where rates dropped down to 3%, but then the years corresponding to his successor (Cristina Fernandez de Kirchner, 2007–2015) saw vacancy rates increasing again as the populist and anti-globalisation policies implemented by her administration clearly had an effect by detracting and deterring business activity. Vacancy rates during her term reached 9% and were reduced by 3 percentage points when the next more liberal administration took office in December 2015.

Brazil's available data ranges from 2005 to 2016. In both cases, São Paulo and Rio de Janeiro show that the local office market suffered from the drop in oil prices happening around 2008 when the world was also facing the international financial crisis. This can be seen in Rio de Janeiro where vacancy reached 11.1%. The unfortunate coincidence of the World Football Cup (2014) and Olympic Games in Rio (2015) with the corruption scandals, has clearly affected the rental office market. The hosting of international events led to regeneration projects (for example, the entire regeneration of Porto Maravilha in Rio), which injected a new supply of prime office space. However the impeachment and corruption scandals considerably reduced business confidence and according to the database Buildings,[6] new office space in the pipeline has been put on hold. The combination of both scenarios (and influx of new office space supply and a reduction in businesses) produced a hike in vacancy rates for class A space in both cities. Rio de Janeiro is the most affected, showing a rate of 40.2% in 2016 – this hike can be linked to the Petrobras scandal as well as overbuilding – while São Paulo has a rate of 23.4% for the same year, a hike that in this city can be attributed to overbuilding.

The next largest longitudinal series of available office data is that of Bogota, which ranges from 2010 to 2016. Undoubtedly, the steady path of the Peace Process that began in 2012 and ended with the signature of an Agreement in October 2016 is boosting business confidence in the country. The increase of vacancy rates in 2015 to 7.8% is related to a supply of nearly 400,000 square metres of class A office space that new developments such as those located along the Calle 72.

Data available for Santiago's office market is from 2012 to 2016. The steady increase is also connected to new supply as the regeneration project of Costanera Centre included the Gran Torre Santiago, which was completed in 2014. The shortest longitudinal series is that of Lima, which ranges from 2013 to 2016. As with Chile and Colombia, the hike in 2015 is connected with new supply, in the case of Lima completed in the neighbourhoods of Miraflores but also San Isidro, the new financial district of the city.

iii) Net absorption

As stated before, absorption and net absorption are also important indicators of the state of a rental office market. While the first one shows the floor space with new rented agreements in a period of time (usually a year), the second one takes into account overdue contracts, demolitions, or removal of space from the market over the same period of time.

The net absorption will therefore tend to be negative if the local economy is shrinking. A negative value means there are changes in the market – for example, companies providing notice of termination of contract to landlords and intending to move to smaller premises. In this case, the surplus square metres not occupied by these companies will appear as overdue contracts while the new space taken by these companies will be deducted from the vacant space at the end of the period under analysis. Equally, companies closing down altogether will also add square metres to the vacancy at the end of the period. The net absorption gives therefore a more detailed indication of what is happening in the job market.

Graphic 2.4 presents the net absorption for grade A space as reported by Cushman and Wakefield for the cities under study here for the same periods as indicated in the previous section. It can be seen that the office rental market in Lima and Santiago de Chile have a positive net absorption and an upwards inclination of the curve. Indeed in a one year period from 2014 to 2015, Santiago's office market had a net absorption increase of +30,000 square metres of office space while Lima's was of +90,000 for the same period. Compared to the change in absorption for the same period in São Paulo (−37,462 sq. m), the problem is that in this case the resulting net absorption is negative. Furthermore, the curve in Graphic 2.4 show that São Paulo's net absorption has been declining since its peak in 2013, with some periods of recovery, certainly, but these have not been sufficient to revert the overall trend. The same can be said for the office market in Rio de Janeiro, which has been on a steady decline since 2010. Bogota and Buenos Aires also show a decrease in net absorption for the period 2014–15.

Considering that new supply has been added to all cities (Graphic 2.2), Lima and Santiago are showing that they have the capacity to absorb the new demand.

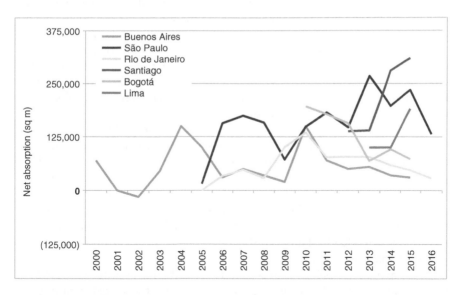

Graphic 2.4 Net absorption

Indeed they have the lowest unemployment rates at country level with 6.6% for Chile and 4.4% for Peru (see Table 1.9 in previous chapter). Argentina and Colombia have an unemployment rate of just under 10% while Brazil is over 11%. Colombia's political stability can help but unemployment levels are forecasted to stay at the same rate at least until 2019. The declining net absorption rate should be considered as a sign that developers and investors need to be cautious in their decisions in Colombia. Brazil and Argentina do not have the political advantage of Colombia. Furthermore, corruption scandals in Brazil are escalating and will certainly affect Argentina's job market as the two countries are close business partners via the Mercosur agreement.

2.3 Shopping centres in South America

As in the previous section that covered the office market, this part of the chapter will now review shopping centres, focusing on their main characteristics in terms of location and generation of revenues (2.3.1); principal components in lease contrasts (2.3.2) and an analysis of the market conditions in South America (2.3.3). The final section (2.4) presents the conclusion to this chapter for all commercial space: office and retail.

2.3.1 Classification of shopping centres and generation of revenues

According to Lambert (2006), shopping centres can be classified into two formats: traditional and specialised ones. The traditional centres are designed to satisfy regular consumers' needs and therefore hold general goods. The specialised ones are focused on non-regular consumers' needs and hold items that demand a detail search on behalf of the consumer in terms of quality and price.

It can be deducted from the above that in order to cater for comparative goods, traditional shopping malls require a larger retail floor area in order to shelter more shops to widen customers' choices. Thus, Lambert (2006) further classified traditional shopping centres by area, from the very large ones (80,000 sq. m or more) to the smallest ones (5,000 sq. m).

Some examples of specialised shopping centres are, according to this author, factory outlets, retail parks, or theme-oriented shopping malls. As customers do not search for specialised goods or services on a regular basis, specialised shopping spaces tend to be located in more remote locations of the city, as customer's journeys to it are not made on a regular basis. Conversely, traditional shopping centres catering for general consumer goods demand more frequent visits from customers and are therefore more centrally located. Although Lambert's classification is intended for European shopping centres, it can be applied to South American markets.

Additional to shopping centres, another trend that continues to grow in importance in Latin American cities as a whole are the mixed used projects (Lizan, 2015). The author attributes this fact to the high cost of land and the fact that customers are looking for convenience and trying to reduce travelling times. In São Paulo, for instance, the recent revision of its master plan (2014) brought some incentives for new mixed use developments that include retail and residential space. Some examples are Cidade Jardim in São Paulo and Costanera Centre in Santiago.

Rental agreements in shopping centres are usually a mix between two rental models: a fixed rent and a percentage rent model. Percentage rent model means that a fixed percentage is applied to the store revenues (known as the variable rent). The calculation of this variable rent is based on the monthly sales of the shop and the kind of retail activity allowed in that particular location inside the mall. The fixed rent model, on the other hand, is a monthly rent (known as fixed-fee), and is not linked to the shop's performance. The difference between the two amounts, the variable rent and the fixed-fee, is called the *overage*. This double rent model is acceptable in most developed countries and also widely accepted in South America. The minimum rent has two different objectives. On one side, it allows the developer/owner to mitigate losses during periods of slow trade due to the retail seasonal cycle (for instance a stationary outlet outside school term times). On the other side, it encourages more transparency in the reporting of turnovers by tenants. Notwithstanding, many disputes arise particularly in times when the retail environment is tough with low sales volumes. This can trigger claims and disputes between landlords and tenants with the latter lobbing against the former in order to achieve most favourable rental terms. Because of this, some shopping centre managers in Brazil have resorted to monitor sales of each shop, although this can be effective in increasing transparency in the declaration of turnover, it may not contribute to the building of a friendly relation between the two parties.

This kind of rental agreement implies that the shopping centre has, in fact, two sources of revenues: the consumers that spend money in the stores and thus contribute to the variable rent, and the shop tenant who pays the fixed rent. The estimated overage will only be achieved by the owner of the shopping centre when each store reaches the expected sales performance used in the calculation during the design stage of the shopping. Of course, this performance needs to be supported by a good management of the shopping centre in order to attend to consumers' needs in regards to the availability of goods and price range as well as other elements to improve the shopping experience (for instance in terms of comfort at the arrival to mall, transport accessibility, availability of parking facilities, and the quality of the food court).

2.3.2 Lease contracts

Shopping centres' contracts in South America usually follow rental terms already observed in the office markets in section 2.1 of this chapter. Tenant contracts for shopping malls are made for pre-defined periods in which the developer/owner commits to the tenant the right-to-use of the *goodwill*. This right-to-use can be charged (known as allowance or *res speratae* clause in the contract). Either the fix or the variable part of the rent as well as the *res speratae* payment can vary according to the tenant. Indeed the importance of having a particular well-known brand in the shopping centre, the location of the store inside the mall, the display area required by the retailer, as well as any other specific demands made by the tenants, are considered when defining the terms of the contracts and charges.

Besides the commitment to paying the rents, the tenants also pay a service charge to contribute to the running expenses, which includes building insurance; electricity; and general maintenance and cleaning. Tenants as well as the owner of the shopping centre contribute to a marketing fund, which is an additional fee to pay for promotion activities related to the mall.

During annual retail cycles, the rent charged might be doubled by some managers as higher volumes of public are expected in malls during these months. This increment is legally explained by the variable rent mentioned above. The minimum rent can also be doubled during these times, as the number of customers rise and therefore shopping centres' services can be in greater demand.

The Gross Leasable Area (GLA) that appears in the contracts comprises the shop floor and sometimes, depending on the design of the mall, a stock area small enough to just provide for daily restocking. As the shopping centres areas are expensive for the shops, they usually do not have a large stock area inside the malls.

2.3.3 Shopping centre markets in South America

The most representative shopping centre markets in South America are presented in Graphic 2.5, which shows data published by the International Council of Shopping Centres (ICSC) for the region for 2014 including forecasted market growth until 2025.

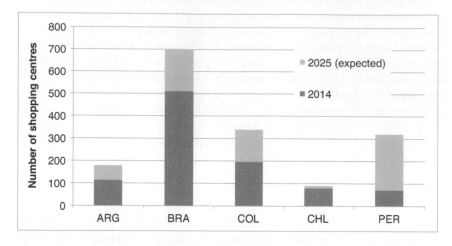

Graphic 2.5 Shopping centres in South America

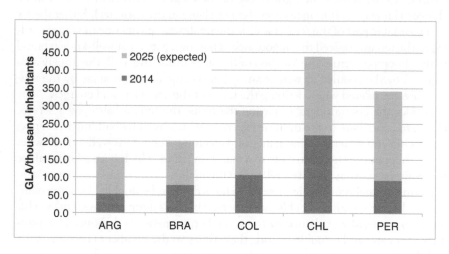

Graphic 2.6 GLA/thousand inhabitants per country

As the graphic indicates, Brazil is the largest market in the region, with 511 (in 2014) malls followed by Colombia (196); Argentina (114); Chile (79) and Peru (71). The market that is expected to have a large expansion is Peru, which is expected to increase the number of shopping centres to 249 by 2025.

The variable GLA/thousand inhabitants in Graphic 2.6 shows the density of shopping malls or the available supply of space in relation to the country's population and is an important indication to study markets and evaluate the need for future developments. Data for Graphic 2.6 has been compiled using data published by the ICSC for 2014 including forecast up to the year 2025.

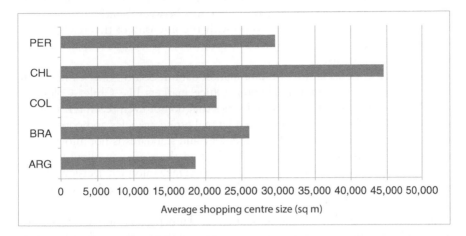

Graphic 2.7 Average shopping centre sizes per country

As the graphic indicates, Chile is the best served in terms of shopping GLA/
thousand inhabitants, although it also has the smallest population in the region.
On the opposite side, Brazil, with the largest population, also has the largest
number of developments. Notwithstanding, the country still shows space for sustainable growth, particularly since in recent years sales in shopping centres in the
country account for 19% of Brazilian consumption.

Countries under study here also present differences in relation to the size of
the shopping malls. Graphic 2.7 shows average size per market according to the
ICSC (2014).

As seen in Graphic 2.7, Chile concentrates the biggest developments while
Brazil, Colombia and Argentina present smaller units. The case of Brazil can
be explained by the fact that they are spread in hundreds of different cities,
with more than 50% of them being developed in large cities during the current
century.

In recent years and mainly due to the economic slowdown (see Chapter 1),
many countries in South America are having to cope with decreasing sales in
retail. Some alternative options have arisen, such as pop-up and temporary shops
that are now emerging. This has the benefit that spaces are occupied while tenants do not have to commit to long-term agreements that might be difficult to
fulfil. This can be an alternative way to bring up the occupancy rate in shopping
malls, thus minimising the increase in the costs of common areas.

2.4 Chapter conclusions

This chapter has shown how deeply intertwined are commercial real estate markets with the local politics. This is a link that can be seen in other regions of the

world – for example, the current Brexit negotiations are affecting commercial real estate in the UK.[7] However, the difference between South America and other regions around the globe is to do with the intensity of this relation and how sustained are the impacts suffered by commercial real estate markets when a negative change occurs.

Additionally, the lack of real estate market data in South America has been made evident throughout this chapter. Office markets like Chile, Colombia and Peru have very short longitudinal series that do not even reach ten years of market activity. The expertise provided by real estate agencies and consultants is helping, but their diversity of methodology and standards is not allowing for regional comparisons. More needs to be done in the region to allow for cross country analysis, this will help foreign investors to take informed decisions. Until then, South America will remain an obscure market where local investors and politicians rule the land.

In terms of geographical location of offices and urban expansion, Buenos Aires and São Paulo offer examples of an active local authority that seeks to capitalise from business activities and encourage office development to revitalise urban areas. Santiago de Chile on the other hand, is facing traffic problems as it is failing to control developer's activities. The main difference between these cities is the fact that Buenos Aires and São Paulo's regeneration projects were actively pursued by the local authorities, while Santiago's was driven by the private sector. This can severely affect connectivity of office locations and can create long hours of commute.

Office specifications are connected to this lack of investment in infrastructure, as workers rely on cars for their daily commute. This and lack of connectivity keeps fermenting the development of the fragmented city described in Chapter 1. Poor working conditions at urban level can eventually influence career decisions of office workers who will prefer to find other locations that can offer better quality of life. The risk here is that the high achievers and most talented amongst the workforce are the ones that can relocate to cities that offer better commute and amenities. South American cities need to do more to retain local talent if they want to increase productivity and competition in the global scene.

Notes

1 As reported by Brazilian RE investment company Patrinvest' article. Available at www.patrinvest.com.br/sale-leaseback_post/, accessed on 15th February 2017.
2 https://nextcity.org/features/view/medellins-eternal-spring-social-urbanism-transforms-latin-america
3 www.rutanmedellin.org/es/
4 www.prefeitura.sp.gov.br/cidade/secretarias/urbanismo/sp_urbanismo/operacoes_urbanas/agua_espraiada/index.php?p=19590
5 www.cali.gov.co/informatica/publicaciones/106110/economia_de_cali/
6 www.buildings.com.br/
7 www.ft.com/content/af97167e-9452-11e6-a1dc-bdf38d484582

Bibliography

Archer, W.R. (1981) Determinants of Location for General Purpose Office Firms Within Medium Size Cities. *Journal of American Real Estate and Urban Economics Association*, 9, pp. 283–297.

Baum, A. (2009) *Commercial Real Estate Investment: A Strategic Approach.* Exeter, UK: EG Books.

Bennet, R. (2005) Marketing Policies of Companies in a Cyclical Sector: An Empirical Study of the Construction Industry in the United Kingdom. *Journal of Business & Industrial Marketing*, 20, 3, pp. 118–126.

Brooks, C. and Tsolacos, S. (2001) Linkages Between Property Asset Returns and Interest Rates: Evidence for the UK. *Applied Economics*, 33, 6, pp. 711–719.

Clark, T.N., Lloyd, R., Wong, K., and Jain, P. (2002) Amenities Drive Urban Growth. *Journal of Urban Affairs*, 5, pp. 493–515.

D'arcy, E., McGough, T., and Tsolacos, S. (1997) National Economic Trend. Market Size and City Growth on European Office Rents. *Journal of Property Research*, 14, 4, pp. 297–308.

Dobson, S.M. and Goddard, J.A. (1992) The Determinants of Commercial Property Prices and Rents. *Bulletin of Economic Research*, 44, pp. 301–321.

Florida, R. (2005) *Cities and the Creative Class.* New York: Routledge.

Fuerst, F., McAllister, P., and Murray, C. (2011) Designer Buildings: An Evaluation of the Price Impacts of Signature Architects. *Environment and Planning* A, 43, pp. 166–184.

Geltner, D., Miller, N., Clayton, J., and Eicholtz, P. (2006) *Commercial Real Estate Analysis & Investments.* Eagan, MN: West Group.

Hui, E.C.M. and Yu, K.H. (2006) The Dynamics of Hong Kong's Office Rental Market. *International Journal of Strategic Property Management*, 10, 3, pp. 145–168.

ICSC. (2014) *Reporte de la Industria de Centros Comerciales en América Latina. Grupo de Inteligencia de Mercado del ICSC.* Available at www.icsc.org/latam/uploads/RLA_White_Paper.pdf, accessed on July 2017.

JLL. (2016) *Latin American Office Market Mid-Year Overview.* Available at www.jll.com.tr/turkey/en-gb/research/41/jll-turkey-commercial-real-estate-market-overview-2016-mid-year-report, accessed on July 2017.

LaGrange, R.L., Kenneth, F., and Supancic, M. (1992) Perceived Risk and Fear of Crime: Role of Social and Physical Incivilities. *Journal of Research in Crime and Delinquency*, 29, 3, pp. 311–334.

Lambert, J. (2006) One Step Closer to a Pan-European Shopping Centre Standard. *ICSC Research Review*, 13, 2, pp. 25–40.

Liow, K.H. (2000) The Dynamics of the Singapore Commercial Property Market. *Journal of Property Research*, 17, 4, pp. 279–291.

Lizan, J. (2015) *Mixed-use Developments in Latin America, The View's From the Architect's Standpoint.* Available at www.linkedin.com/pulse/mixed-use-developments-latin-america-view-from-architects-jorge-lizan, accessed on 10th March 2017.

McKinsey Global Institute. (2011) *Building Globally Competitive Cities: The Key to Latin American Growth.* Available at www.mckinsey.com/global-themes/urbanization/building-competitive-cities-key-to-latin-american-growth, accessed on 19th May 2017.

Rocha Lima Jr., J., Alencar, C.T., and Monetti, E. (2016) São Paulo's Office Market: Scenario for the Next Years. *Journal of Financial Management of Property and Construction*, 22, 2, pp. 154–173.

Rosen, K. (1984) Towards a Model of the Office Building Sector. *Journal of the American Real Estate and Urban Economics Association*, 12, 3, pp. 261–269.

Skogan, W. (1986) Fear of Crime and Neighbourhood Change. *Crime and Justice*, 8, pp. 203–229.

Wurtzebach, C.H., Mueller, G.R., and Machi, D. (1991) The Impact of Inflation and Vacancy of Real Estate Returns. *The Journal of Real Estate Research*, 6, 2, pp. 153–168.

Zunino, M. (2006) Power Relations in Urban Decision-making: Neo-Liberalism, 'Techno-Politicians' and Authoritarian Redevelopment in Santiago, Chile. *Urban Studies*, 43, 10, pp. 1825–1846.

3 Residential real estate in South America

3.0 Introduction

Residential property in South America unfolds at a complex crossing of policy framework, economic cycles and social pressure. At the policy level, planning regulations as well as social housing programmes have a direct effect on residential markets, as both factors influence the demand and supply of units. (As seen in Chapter 1, less restrictive planning can encourage new developments while the delivery of social housing programmes increases the supply of housing.) On the other hand, the tendency to boost construction of housing schemes to reduce unemployment is a widely used tool across South America's governments. This is mainly because the region has a large unskilled workforce that tends to find employment in the construction sector (Murray et al., 2015). Finally, a confusing landscape of property rights and land titling defects tend to manifest through social unrest, which puts pressure on governments to act. Since the turn of the new century, states' responses to social pressure has been to legalise land titles and formalise informal developments, a measure mostly instigated and supported by the Inter-American Development Bank (Brakarz, 2010). This process undoubtedly increases the supply of housing for a segment of the population that are in desperate need of a solution. But it can be argued that the upgrading plans can also encourage the growth of the informal housing market, as informal sellers find willing buyers who agree on the transaction based on the expectation that in the future, land titles will be regulated (Perlman, 2016). This author also states that the informal market can sometimes move at lower prices if compared to the formal, but land speculation still exist here and the question remains as to what is the effect that upgrading programmes have on the open housing market and if (and if so when and how) the two markets will converge.

This scenario presents a unique reality of formal and informal housing speculation, which is more or less shared across all countries under study here. This chapter will therefore provide a brief description of the development process for housing, considering the following factors: 3.1) housing needs in South America according to type of deficit; 3.2) housing programmes targeting the quantitative deficit; 3.3) housing programmes targeting the qualitative deficit; and 3.4) housing provision for the open market sector in the context of the planning system. The final section (3.5) presents an analysis of how all these factors interact and the conclusions to this chapter.

3.1 Housing needs in South America

The existing number of dwellings in a country is an important factor that shapes not only the housing market, but also the housing policies designed by governments. Indeed those bidding in the open market to purchase a house, will pay the price they can afford, therefore the allocation of units will be dictated by the purchasing power of individuals. In this sense, and providing there is perfect competition, the market will gradually allocate the current stock of housing according to the distribution of income in a given location (Harvey and Jowsey, 2004). In very unequal societies such as the South American ones (see Chapter 1) where most people live outside the formal economy and even the minimum wages are insufficient to secure adequate housing (Table 1.16), it is the governments' responsibility to develop suitable policies to help those who do not have the means to compete in the open housing market.

Table 3.1 shows the housing need based on current stock for each country under study here. Data has been gathered by the authors from the official data agencies of each country; therefore, some explanation regarding the characteristics of the data are necessary before any analysis can be presented.

First is the concept of minimum housing standards, which is a relative measure that varies from country to country. Housing need for each country presented in

Table 3.1 Housing need per country (millions)

Country	Million	Year	Previous measure	Quantitative	Qualitative
Brazil	6,068,061	2014	5,846,040 (Fundação João Pinheiro, 2014)	n/a	n/a
Colombia	1,647,093	2012	3,828,055 (Censo Dane, 2005)	554,087	1,093,006
Argentina	3,500,000	2017	2,600,000 (INDEC, 2001)	1,500,000	2,000,000
Peru	1,860,692	2007	1,232,999 (ENAHO, 2001, INEI)	389,745	1,470,947
Chile	1,866,109	2015	2,058,832 (Observatorio Habitacional MINVU, 2013)	391,546	1,474,563
UK			Housing demand 225,000–275,000 per year to keep up with population growth (DCLG, 2017)		

Sources: Argentina Subsecretaría de Vivienda y Habitat (estimates 2017). Given data conflict with INDEC, the previous measurement correspond to the year 2001 before changes to the methodology have been introduced.

Brazil Fundação João Pinheiro www.fjp.mg.gov.br/index.php/docman/cei/informativos-cei-eventuais/634-deficit-habitacional-06-09-2016/file

Chile Observatorio Habitacional MINVU www.observatoriohabitacional.cl/opensite_20080317172111.aspx

Colombia DANE www.dane.gov.co/index.php/estadisticas-por-tema/pobreza-y-condiciones-de-vida/deficit-de-vivienda

Peru INEI www.inei.gob.pe/media/MenuRecursivo/publicaciones_digitales/Est/Lib0868/libro.pdf

Data: Assembled by authors directly from government's national data agencies

Table 3.1 is, by definition, the measurement of the extent to which existing stock falls short of the required minimum standard. Therefore, data must be considered for illustration purposes here as minimum standards vary. Additionally, each country has its own methodology for collecting data and calculating the housing need, although all of them use census returns from their respective national statistics offices as the basic data for calculations. Notwithstanding, some particularities arise, for example, in the case of Brazil, where the deficit is based on four main components: a) those living in precarious conditions, b) those in cohabitation living in overcrowding conditions, c) those who are renting and when the value of the rent exceeds 30% of their income; d) those who are renting and are cohabiting in overcrowding conditions (Fundação João Pinheiro, 2016). As this report explains, the first category only includes informal settlements in urban locations, the rest can include a diverse group that can encompass the offspring of the middle classes that are unable to leave the family home, as well as those who are forced to cohabit after taking a downturn or migrating from one location to another. The last two categories (c and d) only refer to those in rental accommodation. Brazil, therefore, does not distinguish between qualitative and quantitative aspects of the shortage of houses, grouping the deficit only by the four categories described above. The advice from the Fundação João Pinheiro, who is responsible for providing the Brazilian Government with accurate information on housing deficits, is that it is up to the State to make decisions as to when to build new (therefore respond to the qualitative deficit) and where to improve the current stock. How the Government makes such decisions is not clear from the report (Fundação João Pinheiro, 2016). However, the tendency in Brazil is to favour new construction as this helps to reduce unemployment (Murray and Clapham, 2015). This is posing a risk, as the country is now facing a surge in the vacancy rate of units delivered during the construction boom of the World Cup in 2014 and the Olympics held in Rio de Janeiro in 2016 (Wittger, 2017).

Unlike Brazil, the rest of the countries under study in this book make a clear distinction between quantitative and qualitative deficit. Furthermore, countries such as Chile have an even deeper segmentation for each category; for example, the quantitative demand is the aggregate of: a) irrecoverable stock; b) overcrowding; and c) severely overcrowded conditions. On the other hand, the qualitative demand includes those domiciles in: a) need of repair; b) need of an extension; and c) need of improved sanitary conditions. Additionally, as the qualitative need can encompass more than one of the above mentioned categories, the Chilean statistics' office makes a further ranking of housing needs according to how many categories they fall into. For example, a house might simultaneously need repairs and improved sanitation, while others might need all three or just only one. This level of understanding of the data is what has allowed Chile to make detailed policies based on prioritisation of needs, and might help explain the dramatic reduction of its housing needs in less than a decade while the other countries continue to struggle (Tables 3.2 and 3.3).

Still, as seen in Table 3.1, all countries have managed a reduction of the housing deficit since the previous measure. Good policy certainly helps but equally important is the amount of resources that a country commits to the implementation of

Table 3.2 Chile quantitative deficit

Year	Qualitative deficit			Total
	Improvement	Extension	Sanitation	
1996	944,422	486,320	531,834	1,962,576
1998	972,103	456,629	510,680	1,939,412
2000	930,685	394,980	423,457	1,749,122
2003	1,104,820	372,568	407,764	1,885,152
2006	1,239,030	334,108	331,624	1,904,762
2009	1,281,542	300,984	269,446	1,851,972
2011	1,321,600	276,343	359,997	1,957,940
2013	1,100,859	250,250	248,376	1,599,485
2015	1,096,202	199,407	178,954	1,474,563

Sources: Chile Observatorio Habitacional MINVU, www.observatoriohabitacional.cl/opensite_20080317172111.aspx

Data: Assembled by authors directly from government's national data agencies

Table 3.3 Chile qualitative deficit

Year	Quantitative deficit			Total
	Irrecoverable stock	Overcrowded	Severely overcrowded	
1996	197,128	131,289	235,571	563,988
1998	157,691	107,169	232,152	497,012
2000	161,934	76,110	236,331	474,375
2003	128,220	85,638	221,743	435,601
2006	79,482	125,882	229,973	435,337
2009	50,290	182,557	214,712	447,559
2011	67,188	227,556	198,611	493,355
2013	31,523	233,274	194,550	459,347
2015	38,904	183,533	169,109	391,546

Sources: Chile Observatorio Habitacional MINVU, www.observatoriohabitacional.cl/opensite_20080317172111.aspx

Data: Assembled by authors directly from government's national data agencies

the housing programmes. The following section explains the most common housing programmes implemented across all countries and presents a chart indicating the resources committed by each one of them.

3.2 Housing programmes targeting the quantitative deficit

Government intervention to secure adequate housing for those that cannot compete in the open market usually addresses the qualitative and the quantitative deficit. Policies addressing the quantitative deficit imply new construction that can be delivered by the state for rent or ownership. Another alternative also delivered with the help of the state is progressive building, which is only available for ownership. Policies addressing the qualitative deficit include a variety of products such as individual home improvements, neighbourhood upgrading programmes, and regularisation of titling issues. Regardless of the product, all qualitative alternatives are for home ownership. Figure 3.1 summarises all the options.

Home ownership has been encouraged by governments in all the countries, with state subsidies being primarily directed towards this tenure. Direct provision of housing is usually sold to the new occupants through the provision of state subsidies and loans that are sometimes provided by the private sector. The subsidy system is similar across all countries studied in this volume and follows the A-B-C system for its Spanish acronym for savings, subsidy and credit (Ahorro – Bono – Crédito). Most countries allocate subsidies on a means tested basis and distribute available resources accordingly. All of the programmes described in Figure 3.1 have a certain degree of subsidies, which are progressively reduced as households' incomes increase. In the social pyramid of wage distributions, where higher earners are at the top and those on a minimum wage are at the bottom, housing solutions for the top section of the pyramid will have no involvement of the public sector – and therefore receive no subsidies – while the private sector solutions (such as those offered by the banks in the form of mortgages) become considerably more important. It is worth mentioning here that Argentina is the only country of the Big Five where mortgage solutions for the open market remain restricted due to high levels of inflation. Chapter 4 provides a more detail analysis of the financial options available in this country. Figure 3.2

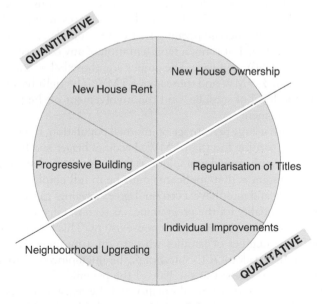

Figure 3.1 Social housing solutions according to deficit type

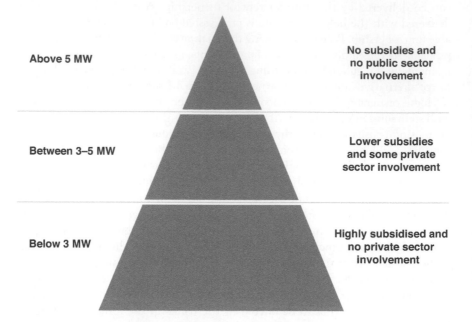

Figure 3.2 Subsidies and wage distribution

illustrates the schemes according to the most representative segmentation of the market for all countries under study here, taking into consideration minimum wages (MW) per household. The graphic is a representation of the most common division of the income pyramid that can be found across all countries under study here, although it is not a representation of any country in particular and cut off points of MW can vary. The groups are: those below three times the MW; those between 3–5 MW and those above 5 MW. The right hand side of the graphic shows the level of subsidies and the involvement of the private sector available to each segment.

For countries with a large percentage of informal population, the bottom of the pyramid (i.e. those earning less than 3 MW) becomes bigger and burdensome to the state, who is the only providor of solutions for this group. Indeed, the problem with the A-B-C system is that applicants must be in full employment in order to access the subsidy and must have a certain degree of savings capacity in order to gather the funds to qualify for the programme. As seen in Chapter 1, the saving capacity for those on a minimum wage is non-existent. This low level of income leaves a significant number of families who are on a minimum wage without a housing solution. In addition to this group, the unemployed as well as those in the informal market coexist at the base of the pyramid. Table 3.4 shows the combined percentages of those in informality and the registered as unemployed per country under study. As stated in Chapter 1, the levels of informality are

Table 3.4 Population outside the ABC system

	Table 10 (average calculated by authors)	Table 9 (2017)	Total % of population outside the ABC system
Argentina	32.9	8.47	41.37
Bolivia	71.8	4.00	75.80
Brazil	32.65	11.54	44.19
Chile	32.1	7.64	39.74
Colombia	51.85	9.60	61.45
Ecuador	45.55	6.90	52.45
Paraguay	66.35	5.47	71.82
Peru	56.2	6.00	62.20
Uruguay	30.45	8.51	38.96
Venezuela	45.25	21.38	66.63

Source: Authors' own from Chapter 1

estimates only and some of them might or might not be registered as unemployed, so details here are for illustration only. The table shows that most countries are failing up to 40% of their population. The only two countries under this mark are Chile and Uruguay, but only just, with 39.74% and 38.96%, respectively. The worst examples are Bolivia and Paraguay, where over 70% of the population are unable to access a house via the ABC system. Colombia, Peru and Venezuela are the next worst performers, failing over 60% of their citizens. The situation of Peru is even worse considering that, as seen in Chapter 1, a household needs three minimum wages to survive, making the bottom of the pyramid even larger for this country. It is evident that policies are failing in all countries, given that a large percentage of the population are having to find their own housing solutions by living in informal accommodation.

For this reason and since the turn of the new Millennium, most countries in the region are now targeting the very poor. The Colombian government launched the *100,000 Viviendas Gratis* (free housing) plan in 2012 (Gobierno de Colombia, Haciendo Casas Cambiamos Vidas).[1] The beneficiaries were families that suffered from environmental disasters or that were displaced during the drug trafficking wars. This new scheme was a last attempt to target the elusive bottom of the pyramid, i.e. the very poor families, and follows World Bank recommendations given to Colombia. This is certainly a departure from the World Bank previous recommendations during the 1980s, when free housing provision was seen as dangerous as it could induce larger rural-urban migration, straining the cities even more (Pugh, 1990). The fact that most South American cities have now large urban populations (Uruguay is, for example, the most urbanised country in the world; see Chapter 1), might explain the change of policy recommendation provided by the Bank to regional governments.

In Brazil, the Growth Acceleration Programme (or PAC for its Portuguese acronym), was launched by former president Ignacio Lula da Silva in 2007 and continued by its successor the impeached Dilma Rouseff. The programme covers different areas of urban development including infrastructure and urban cohesion, but one of the areas relevant for this chapter was the housing investment prioritised in the PAC and delivered via the Programme Minha Casa Minha Vida (PMCMV).[2] As with other social housing plans explained here, the programme targeted those at the base of the social pyramid (earning below 3 MW and up to a maximum of 10 MW with subsidies diminishing as household's salaries increased). The financial mechanisms behind this programme are explained in Chapter 4, while the results of the PAC are provided in Expert box 5.2. But it is worth pointing out here that the PMCMV had some success in delivering homes and in improving informal developments (see Case Study 3.1 in this chapter), but also there are some who point out some failures. For instance, there are claims that the Programme was a way the government faced the 2008 financial crisis and that it is mainly driven by the construction sector (Arantes and Fix, 2009; Macedo, 2010; Cardoso and Leal, 2010). There are also problems with quality (Kowaltowslki et al., 2006) with some critics stating that if customers had a choice by being able to buy a non-subsidised home, the whole scheme and the construction companies that benefitted from it would be out of business (Formoso et al., 2011).

After the default in 2001, Argentinian construction companies restructured considerably. But in 2012 the National Government launched a new plan (Plan Pro-Cre-Ar)[3] that offers 400,000 subsidised mortgages to working families. The first phase of the plan ran until 2014. The subsidy is for progressive building or new homes. In common with similar schemes in other countries, the plan aims to increase growth in the economy through the construction sector, but unlike Colombia's new homes, Pro-Cre-Ar still misses the bottom of the pyramid as it only targets those in employment and/or with savings capacity. Furthermore, the plan relies on the national state, as well as provinces and municipalities to make land available for new developments. A measure that strains already cash-poor municipal authorities, which can only comply by providing low-cost peripheral land rather than in the centre of the cities, where accessibility to employment and other benefits cities can offer are more widely available. The current government, which took office in 2015, has made some changes to this programme. For example, they have changed the allocation system, implementing a means-testing assessment process for those aspiring to receive a new home. However, in a recent presentation at Habitat III by the Secretariat of Habitat and Housing of Argentina, the country stated that the bottom of the market for social housing is still a problem.[4] Another alternative the Argentine government is currently pursuing attempts to bring private sector involvement in the development of housing, a scheme regulated by law (Ley 27.358 for PPPs). In order to entice investors into the social housing sector, the Government is introducing Conjuntos Urbanos Integrados (CUI), which are developments that are promising to deliver social mix by offering open market, affordable and social housing units in one single project. Private partners can join the government with the provision of land, or can provide financing while the Government provides the land by using brownfield plots, which are currently held by the National Agency of Private Property.

The developments share the infrastructure and while the private sector ensures a profitable business from the open market sales of units, the Government receives at least 20% of the total units for social housing purposes.

The CUIs are certainly an innovation in housing policies in South America and only time can tell their success. Still, using state owned land for housing programmes can have its problems. Neighbouring Chile, for example, a country that pioneered the subsidy system, also relies on state land for its current MINVU programmes. However, rising land prices and the failure of the system to impose stricter regulations on developers means that the Chilean state is the main provider of the sites for social housing (Brain and Sabatini, 2006). As these authors suggest, increased construction costs are also part of the problem, and given that the subsidies have a cap, the only solution to keep developing social housing is by building on state land. This makes the authors speculate on the possibility that the state is undervaluing sites in order to complete politically charged projects to win more votes.

3.3 Housing programmes targeting the qualitative deficit

As shown in Figure 3.2, the qualitative deficit is targeted via individual home improvements, neighbourhood upgrading programmes, and regularisation of titling issues. This is a complete change from previous decades (1970s and 80s) when slum clearance was the prevalent solution for dealing with informality (Murray and Clapham, 2015). The change is possibly the result of the region being one of the most socially active if compared with other emerging markets. Community groups, NGOs and other forms of social activism have erupted after the transition to democracy during the mid-1980s, with countries like Brazil and Chile now held up as examples for their achievements in participatory budget and urban upgrading programmes. Not surprising, the headquarters of the World Social Forum is in Porto Alegre, Brazil. Paradoxically, these social approaches are emerging in parallel to a more neoliberal approach taken by some municipalities which are viewing the upgrading of informal developments not only as development opportunity for the private sector, but also as a means to legalise land titles and turn slums into taxable zones.

The tendency to upgrade existing stock in informal developments was first presented to a Western audience by the British architect John Turner in the 1960s.[5] Turner had been working in Peru since the 1950s and 1960s, and was able to witness and take part in a series of progressive experiments and debates that aimed at finding new solutions to informal developments known in the country as *barriadas* (Bromley, 2003). At the time, Peruvian thinkers and international academics begun studying the characteristic tendency among Peruvian rural migrants to settle on urban plots and gradually develop their homes as they gathered materials and resources. During the 1970s the Peruvian government supported this tendency by issuing new legislation to regularise land titles (Leyes de Barriadas). This measure was followed in the 1980s by the creation of a bank to offer building materials credit (BanMat) and culminated in the 1990s with the Commission for the Official Registration of Informal Property (COFOPRI) during the administration of Alberto Fujimori (Fernandez-Maldonado and Bredenoord, 2010).

Although the concept of progressive building seems to have originated in Peru, it was firstly introduced by the Spanish during the colonies and particularly during the great urban expansion of the XVIII century (Murray, 2008). In colonial Latin America, most Spaniards arriving from Europe and seeking better opportunities in the New World travelled in stages, migrating first the male head of the family, who after securing a plot of land from the Crown, began building the family home in a progressive fashion, building a minimum room for one person and then extending and adding more rooms as the rest of family members were arriving. The practice is deeply rooted in Latin America's society and as a result it is widely used in the region. As a consequence of this continuous tendency to rely on progressive building, most countries currently have a large qualitative housing deficit rather than a quantitative one (see Table 3.1). Targeting the qualitative deficit attracts other actors, such as international aid organisations, NGO's and charities, which are filling a crucial gap left by governments' failing policies. The range of solutions are wide from the self-help programmes supported with micro financing options to cover for building materials, such as the Paso a Paso programme implemented in Ecuador since 2001 by the non-profit CIUDAD; to niche solutions for home improvements for families who have to make home adaptations for a family member with special needs, such is the Colombian programme "Improvement without Barriers" implemented in Medellín and providing subsidies to households who are below the two minimum wages.[6]

In terms of neighbourhood upgrading programmes, the most common schemes can take three different formats: a) temporary relocation of families and reconstruction of houses; b) the improvement of the areas' infrastructure (bringing in sanitation, water and electricity); or c) a mix of the two; that is, infrastructure plus some reconstruction of houses. Strategies depend on geographical conditions in each location, and decisions are taken according to environmental safety. The examples presented here illustrate two possibilities: Quinta Monroy (Box 3.1) was an informal development whose settlers each received a new dwelling, while Cantinho do Céu has seen infrastructure improvements with some relocation of families that were inhabited in hazardous locations. In both cases, the preferred tenure is ownership and land titles were regularised during the process.

Solving titling issues is the third and final option for addressing qualitative deficits in South America. Again, it was Peru who pioneered the regularisation of informal plots of land by introducing a new legislation in 1996 (Ley de Promoción del Acceso a la Propiedad Formal, Decree 803, 22 March 1996). The law established COFOPRI (the Comisión de Formalización de la Propiedad Informal), which had functions to oversee and legally register informal properties to the Registro Predial Urbano (cadastral registry). Since then, and according to the World Bank, COFOPRI has legalised nearly 1.5 million properties and has expanded its actions according to changing circumstances; for example, the provinces affected by the 2007 earthquake were added to COFOPRI's list of regions (Ley 29802). The success of the organisation has added functions to its remit and is now tasked with identifying vacant land owned by the state that might be suitable for progressive building (Decreto 1202). COFOPRI has now title holding to all government lands, whether fiscal or municipal, similar to the Argentine and Chilean examples described above. Another important regulation was issued in 2015 (Decreto Supremo 009-Vivienda) by which COFOPRI is exempt of land registry payments at national level. This means that the process of regularisation of titles has no impact on the families who are looking to formalise their properties through this programme. These series of regulations and decrees have provided the organisation with the independence and administrative autonomy needed to deal with the problem while operating within the national framework of institutions (for example they are still dependant on Central Government funding and they respond to the Ministry of Economy where they submit annual reports of accounts).

Additionally, the system now has a series of manuals to guide applicants through the process (Manuales de Procedimientos – MAPRO), which cover individual as well as community projects, they also have a system to train local professionals to support the administrative activities of the organisation (COFOPRI, 2015). This system of standardisation and expansion of trained consultants have been praised by organisations such as the World Bank and the IMF. These institutions support the views of Hernando de Soto in relation to land titling problems being a barrier for the poor to access to capital by using their assets as collateral. But this is a contested ground that needs further research. A study funded by the Norwegian Government (Durand-Lasserve et al., 2007), which carried out an extensive literature review on the topic, concludes that there is little evidence to demonstrate the economic benefits to reduce poverty by facilitating access to credit markets. Furthermore, as pointed out at the start of this chapter, questions remain on the effect of mass titling programmes and the distortion of domestic land markets, as well as the impact on foreign and home investment. On the positive side, some of the administrative and legal challenges presented by the report have found a solution in Peru by the creation of COFOPRI, mainly in relation to the exception of registration costs and the freedom the organisation has within the national government (see above). There is also an incipient connectivity between land registration and urban planning through new mapping tools, but the fact remains that social demand is still driving urban growth and this has serious consequences to the environment.

3.4 Housing provision for the open market in the context of the planning system

As seen in Chapter 1, the income distribution in South American countries is very unequal, which puts more pressure on governments to design and implement corrective policies. In addition to these market inequalities, it is also necessary to understand the planning system in the context of regional patterns of behaviour that are deeply rooted in the region's culture. Indeed, in South America, there is a natural tendency of the social classes to drift towards the same areas, thus creating enclaves of extreme wealth or poverty. As housing policies are needed to correct inequalities arising by the inability of certain groups to access the open market, planning policies are also needed to correct this tendency of social agglomeration and to break those patterns of geographical inequality in order to guarantee that all citizens can benefit from the advantages of urban life. South America has systematically failed in designing and implementing corrective policies to address this problem, and this is mainly because of the prevalence of two main evils: gated communities and weak municipal authorities.

Gated communities have been defined as residential areas were public spaces have been privatised; they can either be in the city or the suburbs and in affluent or poor neighbourhoods (Blakely and Snyder, 1997). The mode of transport prevailing for the gated communities is the car, following the American example of peripheral cities, which was encouraged through the New Urbanism charter. In all countries under study in this volume, urban sprawl accelerated when motorways were expanded during the 1990s, fuelled by private investment. As motorways are mainly private enterprises, most operate on toll systems which, for those who can afford it, considerably reduces the commute to the city (Thuillier, 2005; de Duren, 2006; Borsdorf et al., 2008). Inevitably, shorter commutes increased the demand for dream houses in suburbia.

The families relocating to gated communities are mostly of the same socio-economic background, as the level of service charge for maintenance of communal grounds and facilities (which can be as diverse as swimming pools to golf courses and helipads), act as a barrier to social mix. The newest trend is "affordable" condominiums for the middle classes, which have fewer amenities but still enjoy private security; the latter, one of the main concerns for many in South America when looking to buy a house (see crime perception table in Chapter 2 and also Coy and Pöhler, 2002; de Duren, 2006). Infrastructure around these complexes usually includes health centres, education facilities and cemeteries, all privately paid by local residents. New businesses arriving to the area either follow or precede the opening of a gated community, including large shopping malls and supermarkets (Coy and Pöhler, 2002; Borsdorf et al., 2008). From the 1990s onwards, and given that most municipalities lacked a long-term plan, most of these infrastructure projects are executed by private investors, who are mostly responsible for the current fragmented layout of the periphery of most cities across South America. Briefly and broadly speaking, all countries share the same pattern: the new community is initiated by a developer or a business consortium who assumes the responsibility for the entire project, from gathering investors, to buying the land, architectural design, planning application, construction and sales of units to final consumers. After the families move in, services and maintenance of the communities' site can be provided by the developing firm, outsourced or can be undertaken by the residents themselves organised in a cooperative (Coy and Pöhler, 2002; Borsdorf et al., 2008).The urban periphery is the preferred place for most gated communities, sometimes encroaching agricultural lands. This is due to land availability and also given the high return that can be achieved compared to more consolidated areas of the city. Turning rural land into urban, for example, reportedly generates profits of up to 550% of the original value (Chiape de Villa, 1999).

Some argue that these types of developments bring benefits to areas that are characterised by informality (Sabattini et al., 2001; Salcedo and Torres, 2004). The argument is that the new spaces of the wealthy are providing workplaces for the poor, for example, in construction, security, and domestic service. Figures published so far show that this is partly true (Thuillier, 2005). However, the benefits that municipalities receive in terms of developers' contributions and taxes from new residents have been questioned by some (Brain and Sabatini, 2006; Thibert and Osorio, 2014).These authors maintain that new developments tend to be located in poor municipalities, with a large informal population and little infrastructure. This means small local authorities that are already struggling to collect revenues. Therefore, these offices see the arrival of a new gated community as a way to generate jobs and tend to compete with other municipalities to attract these developments by offering tax incentives to developers. The case of the small municipalities in metropolitan Buenos Aires is paradigmatic. Here the planning process is lengthy and opaque, as planning delay usually pushes developers to break ground as soon as the land is purchased. This is done with the hope that planning permission will be "solved" at some stage by the local officer (de Duren, 2006; Thuillier, 2005). To add to the problem, once the new wealthy residents move in, their tendency is to contest municipal tax as they consider that all their services are privately paid so they see no reason to pay municipal taxes.

Faced with rural to urban migration problems during the late 1990s, local authorities in Colombia were requested to develop a strategy plan (*Ley de desarrollo Territorial 388*, 1997 – still in force) (Alcaldía de Bogota Ley 388, 1997). New municipal funds were also created (*Fondos Municipales de Vivienda de Interés Social*), with the aim of collecting and managing revenues for social housing programmes. While the strategy plan allows municipalities to identify land for social housing, the municipal funds equip them with regulations to pursue development contributions from the private sector. This is done either by the demand of land for social housing from the land owner or by a development levy paid by developers (Chiape de Villa, 1999; Tachópulos Sierra, 2008). However, there are concerns that the regulation is not fully enforced (Carrión Barrero, 2008) and that municipalities only produce a plan just to comply with the national law; and the reality in Colombia is very much like that of Chile and Argentina where developers rule the land.

3.5 Chapter conclusions

Mass social housing programmes in the countries studied in this chapter run parallel with the rise of the *megaproyecto* for the upper classes. These are larger than previously developed communities that are now intended for more than 50,000 people (Coy and Pöhler, 2002; de Duren, 2006). Furthermore, given the increasing fear of crime (Table 2.2), all sectors of the population these days prefer the sense of security offered by gated developments and so the "gated megaproyectos" now serve the wealthy as well as the poor (de Duren, 2006; Borsdorf et al., 2007). The fragmented city is, thus, spreading across the region where gated ghettoes of the upper class are "mixing" with gated social housing projects in the periphery. The monotony of the urban landscape exacerbates when the same consortium of investors and developers (bearing in mind they are also architects and designers) operate across many Latin American countries (Coy and Pöhler, 2002, p. 358). Still, most problematic is the land struggle that this generates: where a new gated community for the wealthy appears, it immediately takes up good land that becomes unaffordable for social housing (Brain and Sabatini, 2006). Land squatting becomes then the only solution for those who seek employment in the new gated community. Informality erupts and spreads around enclaves of wealth, usually taking poor quality land that has no interest for real estate investors. The pattern of a sea of informality, with patches of wealth and housing estates described as the fragmented city, is then completed.

Notwithstanding, some changes are emerging through land taxation and captura de plusvalia (land value capture). Most authors agree that Brazil is leading the way in development contributions. The Special Social Interest Zones (ZEIS) is a new regulation that requires cities to contribute with a certain percentage of social housing for all developments (Prefeitura de São Paulo Decree 44.667, 2004). Currently, this regulation has only been applied in São Paulo but with promising results, given that social housing schemes resulting from these programmes intend to target different sectors of the population, encouraging a more social mix (Budds et al., 2005). The other system used in Brazil is the sale of vouchers for development rights in exchange for social housing. In this system, if a developer wants to build above the ratio that the size and location of the plot allows, he or she can negotiate this and obtain vouchers that can be either auctioned in the real estate market or used in other sites. In addition, properties that are unoccupied and are considered suitable for social housing whose owners have a land tax debt, can exchange the property for land tax credit. The legislation means that there are no monetary transactions, which increase transparency for parties, the debtors and the creditors, and the local authority in question acquires a new unit of RE for social housing.

In contrast, Argentina and Chile still fail to implement planning gain policies, leaving the responsibility to deliver serviced land to municipalities. As a consequence, and given the extremely difficult economic circumstances that Argentina currently faces, most municipalities have reduced their social housing targets (Barreto, 2012). Furthermore, some municipalities are so short of resources that they have started to divert most of the subsidies to the payment of municipal employees. A similar situation of diversion of subsidies to cover for other shortcomings was seen in Colombia during the time of the Instituto de Crédito Territorial (ICT), the State-run social housing provider that operated until the 1990s reform (Tachópulos Sierra, 2008) and during the closing days of the INURBE. The deviation of funding from housing to pay for an administrative system that is unable to cope with a large population in need seems to be a recurrent fault throughout the decades in these countries.

The white elephant in the room is the provision of social housing via the rental sector. The housing association type widely available in other countries, such as the US and the UK, is completely absent in South America. The informal market is the main one that currently caters for the tenants at the bottom of the wealth pyramid. This rental sector is characterised by individual landlords, often living in the same property, and by informal letting practices, despite state controls on rents and tenants' rights in the formal sector. There is very little research on the informal rental sector, but hearsay evidence suggests that some of them pay a hefty price in the hands of unscrupulous landlords who operate completely outside the law. As Scheinsohn and Cabrera (2009) suggest, the only hope for social housing projects are in the hands of NGOs and social activist movements who are able to lobby between the two political factions, the government and the informal landlords. By profiting from the lack of government cooperation, organisations like Madres the Plaza de Mayo and Piqueteros in Argentina are finding a place as social housing developers and providers for the lowest sector.

Case study 3.1 Quinta Monroy and Cantinho do Céu, how to upscale the good examples?

Claudia Murray

Quinta Monroy is located at the very heart of the city of Iquique, Chile. Informal development began there in the 1960s and by the year 2000 there were over 100 families living in precarious conditions in an area of around 5,000 sq. m. Being in the city centre means the land is theoretically high value, and therefore efforts were made to relocate those living there. After several attempts to move families to the edge of the city failed, the authorities decided instead to help the community legalise their settlement and strengthen their roots in the area, appointing the architectural firm Elemental to provide the community with a new housing development.

In 2003 Chile's budget for social housing was very low – just USD 7,500 per dwelling – to cover land acquisition, infrastructure as well as the construction of each house, and yet at the time land alone in the periphery of Iquique was worth around USD 20 per sq. m. This presented Elemental with something of a financial challenge as well as a typological challenge in relation to social housing.

Their strategy for Quinta Monroy was effectively based on sweat equity: a reinforced concrete shell (walls, floors, roof and staircase) is provided, with all the necessary plumbing but minimal fittings. The families themselves then customise their units, adding appliances and furniture at their own expense. The system relies on the families to find the means to finish their houses. Their homes are therefore the materialisation of their earnings in the informal job market. The design team presented the approach to inhabitants and asked them their views and preferences.

The design team presented the financial difficulties to the families and asked them to participate and express their views and preferences for their future homes. In Chile, this was an unprecedented approach to social housing: low income families are typically given little input and just have to accept and be grateful for what they are given. The urban design layout also included four open public piazzas. These spaces were left for the community to customise, such that every family could contribute to the character of the open space, even if this was only by virtue of how they decorated their own façades.

In Brazil, the informal housing issue has been around for much longer, with some of the poorest urban favelas being in São Paulo. The city has been gradually and informally populated since the 1980s, but it was only in this century that the Municipality of São Paulo decided to take action and allocate funds for redeveloping one of the most deprived areas in the city.

Cantinho do Céu lies on the left bank of the Billings dam, which is vital for the city of São Paulo, providing a third of its fresh water. It is also a nature reserve protected by the Brazilian government. Concerns over its potential contamination due to waste disposal in the dam by local inhabitants was a key imperative for the government to take action.

The Municipality commissioned *Boldarini Arquitetura e Urbanismo* to devise a scheme to provide a safer environment for residents and formalise the land titles of their properties. Developing a social housing in the context of a nature reserve was a challenge, even more so because the area has suffered severe flooding in recent years. Most of the building works executed in Cantinho do Céu such as sanitation, flood barriers and road paving were also aimed at recovering and protecting the wildlife. So far the 11,000 families have benefited from the scheme stands and the local wildlife is thriving once again.

The success of both these schemes has been widely publicised in Latin America, with both architectural firms winning awards and Quinta Monroy in particular featuring in an exhibition at the Museum of Modern Art in NYC. The problem that now arises is how to repeat the success at a more wider scale. The question of how to apply *Elemental*'s approach, for example, to one of the largest *favelas* of São Paulo, Paraisópolis, is populated with difficulties. In pursuit of higher density, Paraisópolis resembles European social housing blocks developed in the 1970s and 1980s. Indeed, it can be argued that Brazil has a legacy of modernist schemes and whilst Brazilians are quite familiar with life in tower blocks, their failings are just as apparent as their Europe counterparts.

But it is not just a question of scale in relation to form, the issue is also about scale in relation to process: Quinta Monroy deployed a task force of architects, social workers, engineers and a wide range of academic experts to work on solving the problem on behalf of just 100 families. In comparison the target group in Paraisópolis is 40,000 people, who would probably need to be decanted. In the process of building new units, the Municipality of São Paulo relocated people to temporary camps while bulldozing their current *favela*. The break in continuity raises the issue of whether a community can survive the shock.

The other unintended outcome of formalising settlements is that the inhabitants realise the worth of their new homes by turning landlord: there are media reports of people letting their flats who have themselves moved to another *favela* elsewhere. This tactic has prompted the government to change legislation and unusually only give property titles to women, since they are more likely to value better living conditions and remain with their children in the new units. This brings to mind the question of whether the community that once existed is still effectively intact, given how easily members seem to want to move away after receiving their new houses.

There have been other reports in the media that many new social housing schemes are developed in complete isolation, far from job opportunities and with little transport connectivity. Furthermore, lack of adequate design and poor technical advice are creating new problems, such as proposals to canalise local creeks, which could lead to an increase in summer flooding. Undoubtedly social participation is a step towards equality, but technical supervision needs to be better provided.

In contrast, the intervention in Cantinho do Céu has had less impact on neighbouring areas, perhaps because the architects' contribution was to manage what was already in place. Although some families were relocated, eviction when there are flood risks could not be avoided, and whilst road access has been improved, there are still issues with connectivity as the main transport network of São Paulo does not reach an area that is used by over 10,000 people, but this problem will no doubt be addressed in future.

The bigger issue at stake in South America is now to address the endemic social inequality and lack of a social mobility in urban areas. Even in a small community like Quinta Monroy and despite all efforts from architects and social workers, home owners insist on building fences and gates around their new neighbourhood. The need to demarcate their territory seems too strong to resist, but it has of course produced ghettoes from which social mobility is almost impossible.

No scheme will ever succeed if it is built as a ghetto from the outset, it will simply stand as a symbol of social failure. The ongoing sustainability of current government efforts to reduce informal developments is still uncertain, but what is clear is that current approaches require more conclusive research before governments and policymakers wade in and destroy existing communities and livelihoods.

Notes

1 Gobierno de Colombia, Haciendo Casas Cambiamos Vidas. 100,000 Viviendas Gratis. Available at www.100milviviendasgratis.gov.co/publico/Default.aspx
2 Regulatory Framework of the Programme Can be Found at the Ministry of Cities. Available at www.cidades.gov.br/index.php/minha-casa-minha-vida.html
3 Decree 902/2012 Boletín Oficial. Available at http://procrear.anses.gob.ar/documentos/decreto-procrear.pdf
4 Habitat 3, Presentation by Dr Ivan Kerr, Sub-secretary of Urban Development and Housing, Secretaria de Vivienda y Habitat, Ministerio del Interior, Obras Publicas y Vivienda de la Nacion Argentina. Quito, Ecuador, October 2016.
5 Turner published extensively on social housing policies, sometimes coauthoring with his wife Catherine, the American Anthropologist William Mangin and British Architect Patrick Crooke. For a full list of his publications see DPU-UCL list. Available at www.dpu-associates.net/node/100
6 Both plan reached the shortlist for the UN World Habitat Awards coordinated by the UK charity Building Social Housing Foundation (BSHF). For Ecuador see www.bshf.org/world-habitat-awards/winners-and-finalists/paso-a-paso-strategic-alliances-for-better-housing/ and for Colombia see www.bshf.org/world-habitat-awards/winners-and-finalists/improvement-without-barriers/

Bibliography

Arantes and Fix. (2009) *Minha Casa, Minha Vida', o Pacote Habitacional de Lula*. Available at http://web.observatoriodasmetropoles.net/download/gthab/text_ref_outros/fix_e_arantes_MCMV.pdf, accessed on July 2017.

Barreto, M.A. (2012) Cambios y Continuidades en La Política de Vivienda Argentina (2003–2007). *Cuadernos de Vivienda y Urbanismo*, 5, 9, pp. 12–30.

Blakely, E. and Snyder, M. (1997) *Fortress America: Gated Communities in the United States*. Washington, DC: Brookings Institution Press.

Borsdorf, A., Hidalgo, R., and Sánchez, R. (2008) Model of Urban Development in Latin America: The Gated Communities and Fenced Cities in Metropolitan Areas of Santiago de Chile and Valparaíso. *Cities*, 24, 5, pp. 365–378.

Brain, I. and Sabatini, F. (2006) Relación Entre Mercados de Suelo y Política de Vivienda Social. *ProUrbana*, 4, May Issue, pp. 1–13.

Brakarz, J. (2010) The IDB: 25 Years of Neighbourhood Upgrading. In Rojas, E. (ed.), *Building Cities: Neighbourhood Upgrading and Urban Quality of Life*. New York: Inter-American Development Bank, pp. 141–153.

Bromley, R. (2003) Peru 1957–1977: How Time and Place Influenced John Turner's Ideas on Housing Policy. *Habitat International*, 27, 2, pp. 271–292.

Budds, J., Teixeira, P., and SEHAB. (2005) Ensuring the Right to the City: Pro-poor Housing, Urban Development and Tenure Legalisation in São Paulo, Brazil. *Environment and Urbanisation*, 17, pp. 89–113.

Cardoso, A.L. and Leal, J.A. (2010) Housing Markets in Brazil: Recent Trends and Governmental Responses to the 2008 Crisis. *International Journal of Housing Policy*, 10, 2, pp. 191–208.

Carrion Barrero, G. A. (2008) Debilidades del Nivel Regional en el Ordenamineto Territorial Colombiano. Aproximación desde la Normatividad Política Administrativa y de Usos de Suelo. *Architecture, City and Environment*, 3, 7, pp. 145–166.

Chiape de Villa, M.L. (1999) *La Política de Vivienda de Interés Social en Colombia en Los Noventa*. Santiago de Chile: CEPAL.

COFOPRI. (2015) *Memoria Institucional*. Available at www.cofopri.gob.pe/media/2667/memoria-institucional-2015-cofopri-logo-corregido.pdf, accessed on 31st October 2016.

Coy, M. and Pohler, M.M. (2002) Gated Communities in Latin American Megacities: Case Studies in Brazil and Argentina. *Environment and Planning B, Planning and Design*, 29, pp. 355–370.

de Duren, N. (2006) Planning à la Carte: The Location Patterns of Gated Communities Around Buenos Aires in a Decentralised Planning Context. *International Journal of Urban and Regional Research*, 30, 2, pp. 308–327.

Durand-Lasserve, A., Fernandes, E., Payne, G., and Rakodi, C. (2007) *Social and Economic Impacts of Land Titling Programmes in Urban and Peri-urban Areas: A Review of the Literature*. Available at www.birmingham.ac.uk/Documents/college-social-sciences/government-society/idd/research/social-economic-impacts/social-economic-impacts-literature-review.pdf, accessed on 31st October 2016.

Fernandez-Maldonado, A.M. and Bredenoord, J. (2010) Progressive Housing Approaches in the Current Peruvian Policies. *Habitat International*, 34, pp. 342–350.

Formoso, C., Leite, F., and Miron, L. (2011) Client Requirements Management in Social Housing: A Case Study on the Residential Leasing Programme in Brazil. *Journal of Construction in Developing Countries*, 16, 2, pp. 47–67.

Fundação João Pinheiro. (2016) *Déficit Habitacional no Brazil 2013–2014*. Belo Horizonte: Governo de Minas Gerais.

Harvey, J. and Jowsey, E. (2004) *Urban Land Economics*. London: Palgrave Macmillan.

Kowaltowslki, D., Gomes da Silva, V., Pina, S., Labaki, L., Ruschel, R., and Moreira, D. (2006) Quality of Life and Sustainability Issues as Seen By the Population of Low-Income Housing in the Region of Campinas, Brazil. *Habitat International*, 30, 4, pp. 1100–1114.

Macedo, J. (2010) Methodology Adaptation Across Levels of Development: Applying a US Regional Housing Model to Brazil. *Housing Studies*, 25, 5, pp. 607–624.

Murray, C. (2008) The Regulations of Buenos Aires' Private Architecture During the Late Eighteenth Century. *Architectural History*, 51, pp. 137–160.

Murray, C., Abiko, A., Monetti, E., and Peinado, J. (2015) Research Agenda for the Built Environment in Latin America. *Cuadernos de Vivienda y Urbanismo* (INJAVIU), 8, 16, pp. 226–245.

Murray, C. and Clapham, D. (2015) Housing Policies in Latin America: Overview of the Four Largest Economies. *International Journal of Housing Research*, 15, 3, pp. 347–364.

Perlman, J. (2016) The Formalisation of Informal Real Estate Transactions in Rio's Favelas. In Birch, E., Chataraj, S., and Watcher, S. (eds.), *Slums: How Informal Real Estate Markets Work*. Philadelphia: University of Pennsylvania Press.

Pugh, C. (1990) The World Bank and Housing Policy in Madras. *Journal of Urban Affairs*, 12, 2, pp. 173–196.

Sabattini, F., Cáceres, G., and Cerda, A. (2001) Segregación Residencial en las Principles Ciudades Chilenas: Tendencies de las Tres Últimas Décadas y Posibles Cursos de Acción. *Eure*, 26, 79, pp. 21–42.

Salcedo, R. and Torres, A. (2004) Gated Communities in Santiago: Wall or Frontier? *International Journal of Regional Research*, 28, 1, pp. 27–44.

Scheinsohn, M., and Cabrera, C. (2009) Social Movements and the Production of Housing in Buenos Aires; When Policies are Effective. *Environment and Urbanisation*, 21, pp. 109–125.

Tachópulos Sierra, D. (2008) El Sitema Nacional de Vivienda de Interés Social (1990–2007). In Ceballos Ramos, O. (ed.), *Vivienda Social en Colombia: Una Mirada Desde su Legislación 1918–2005*. Bogota: Editorial Pontificia Universidad Javeriana, pp. 181–238.

Thibert, J. and Osorio, G.A. (2014) Urban Segregation and Metropolitics in Latin America: The Case of Bogota, Colombia. *International Journal of Urban and Regional Research*, 38, 4, pp. 1319–1343.

Thuillier, G. (2005) Gated Communities in the Metropolitan Area of Buenos Aires, Argentina: A Challenge for Town Planning. *Housing Studies*, 20, 2, pp. 255–271.

Wittger, B. (2017) *Squatting in Rio de Janeiro: Constructing Citizenship and Gender From Below*. Cologne, Germany: Urban Studies.

4 Real estate funding in South America

4.0 Financing commercial and residential real estate

South American countries under study in this volume have their respective financial systems at different stages of development, and this is especially true in regards to the spectrum of vehicles and instruments available to fund real estate activities. As it will be shown in this chapter, Brazil is the country that presents a broader range of options, from banking resources to securitised mechanisms, and from REIT-like instruments to covered bonds. Because of this reason, this chapter will focus on Brazilian investment options, but highlighting the equivalent instruments in other countries when appropriate. Another country on focus here is Argentina, as it presents the other side of the spectrum: the nation with the least financial resources for real estate development. Notwithstanding, and even considering that Brazil is the most advanced economy in this sector, real estate credit in general represents less than 10% of Brazilian national GDP, so in the global context this is a region that has a lot of potential for growth in real estate investment.

As seen in the previous chapter, provision of housing has been for decades the role of the estate in many South American countries. However, this role has been changing from supply of units to supporting the demand side of the delivery via a system of subsidies that are in relation to family earnings (i.e. those with the lowest income receive more subsidies, with the subsidised amount diminishing as the families increase their income; see Figure 3.2 in Chapter 3). In order to enter into one of these subsidised schemes, families need to demonstrate their saving capacity; therefore, most families tend to create a savings account, which is primarily dedicated for the purchase of a house. The rest of the funds for home acquisition are provided by banks, which are usually the national mortgage bank with an increasing participation of commercial banks in some countries in recent years.

Still a major barrier to any system of savings for real estate financing is inflation. Most South American countries battle with volatile environments and uncontrolled inflation (see Chapter 1), which makes the availability of long-term debt very difficult. The following sections detail how different countries manage to fund the provision of housing in inflationary contexts, focusing on Brazil (4.1) and including other forms of real estate investment in this country.

The chapter then focuses on residential investment methods in other economies, covering in particular those who imposed a unit of savings system (4.2). Section 4.3 presents the case of Argentina, including public, private and PPP options recently developed by the new administration. The final part of the chapter (4.4) presents the conclusions.

4.1 The case of Brazil

During the mid-1960s, the Brazilian government found itself short of any financial mechanism to support the development of housing and infrastructure, as well as mortgage structures to help with the acquisition of homes. As a result, a system was created in 1964 to deal with this double need of housing and infrastructure. The system, although limited, as will be shown below, continues to this day and is called Sistema Financeiro Habitacional (SFH) Housing Financing System. The SFH has two primary resources: i) the FGTS; and ii) the SBPE.

i) Fundo de Garantía por Tempo de Serviço (FGTS) – guarantee fund for length of service

The FGTS provides resources to support the unemployed and receives 8% of the workers' wages and additional employers contributions that are collected monthly into a fund. This fund is managed by a public bank called Caixa Econômica Federal (CAIXA). The resources collected by the FGTS are sufficient to cover for unemployment benefits, but interest rates received in the accounts are too low (currently 3% per annum), and therefore insufficient to cover for inflationary changes (see Table 1.19). Therefore, CAIXA uses loan charges and other financial income obtained by the Bank to subsidize housing for FGTS' members. This is why the mandate of FGTS also covers provision of infrastructure (mainly sanitation programmes) and loans for home ownership for the lower income sector of the population.

As seen in Chapter 3, financing for the housing sector in Brazil primarily addresses income brackets up to a maximum of 10 minimum wages, and can receive a government subsidy from the National Treasury, such as the Minha Casa Minha Vida Programme. This programme is especially designed for people with incomes below three minimum wages (see Figure 3.2 for comparison). This income group receives a subsidised interest rate plus a government cash payment towards the repayment of the loan; in other words, they are fully subsidised. For those households that are above this income level (i.e. from three to 10 minimum wages), the only support they receive is a subsidised interest rate on the loan.

The FGTS has an annual balance of approximately USD 4.5 billion made up from the difference between deposits and payments to workers' contributions. This constitutes a first set of accounts operated by the FGTS. A second set of accounts comes from the operational management of the fund, resulting from the financial income provided by interests on loans minus the payment to CAIXA, which manages the loans. This second set of accounts amounts to approximately USD 4.1 billion per year of resources. The resident stock of funds considering new loans that are available annually for development and acquisition of housing is a total of USD 14–15 billion.

The board that manages the funds decides on the level of finance that will be available for house financing each year, and the degree of investment aggression in their stance may be influenced by political demands; for instance, the need to boost construction to generate employment (see Chapter 3). The organism that designs the allocation of funds to different programmes is the FGTS Curator Council, composed of 24 members, 12 of whom are appointed by the government, six of whom are representatives of workers unions, and the remaining six are representatives of employers' unions.

ii) Sistema Brasileiro de Poupança e Empréstimo (SBPE) – Brazilian savings and loan system

The second source of funds that feeds the SFH comes from the so-called Brazilian Savings and Loan System (SBPE), which manages the resources of tax-free voluntary savings collected by public and private banks, with a mandate to direct at least 65% of available resources to real estate credit operations, either for development of residential or commercial real estate.

The system as a whole spins a total of about USD 160 billion per year, with USD 104.4 billion in real estate credit and development funding, the latter being primarily intended for the housing segment, which is perceived by banks involved as a lower risk operation. The average savers using these types of accounts do not have access to other instruments that are more profitable, capable of promoting greater wealth gain. This is because they are small accounts that amount to 10% of the cash deposits available in the system. As a result, even if there is a certain potential for growth of this portfolio, the system faces a barrier to expansion due to the limited resources.

4.1.1 Other forms of investment available in Brazil

i) Letra Imobiliária Garantida (LIG) and Letra de Credito Inmobiliario (LCI)

These real estate bonds are an additional source that have helped to grow the supply of instruments to finance commercial and residential real estate in Brazil. The LIG and LCI are fixed income credit bonds issued by banks and backed by real estate operations with a pre-determined maturity. Although the LCI has been fully working for some time, the LIG is currently in the regulatory phase and is expected to be available in 2018. Its format will be similar to the LCI with an additional guarantee from the issuing financial institution, and will follow very closely the model set up by the European Covered Bonds. With the regulation of the LIGs, Brazil is expected to expand the funding instruments even further and meet the needs of developers and investors alike.

ii) Real Estate Receivables Certificates (CRI)

Built like the Mortgage Backed Securities (MBS) and the Commercial Mortgage Backed Securities (CMBS) in the US, CRI appeared in the Brazilian market in 1997.They are certificates that are issued based on financing agreements, leases or any other type of operation that has the property as guarantee of payment. They currently account for about USD 22.8 billion in resources in Brazil. The CRI market still finds some barriers to growth. Perhaps the main issue is the lack of standardisation of the contracts for both the financing and the securitisation process, as these are still executed on an individualised format, which limits the expansion of this vehicle. A similar instrument exists in Chile and is known as Letra Inmobiliaria (LI). The LI are issued by banks according to the required amount and term and immediately sold in the financial market. Credit in LI in Chile can cover up to 100% of the value of the property.

iii) Real Estate Investment Fund (REIT)

FII, for its Portuguese acronym, is the Brazilian structure that was created in 1993 and regulated in 1994. It is similar to the American REITs but the FII market in Brazil still has ample space for development, since as with the CRI, some structural mechanisms for its development are needed. The first of these structural reforms needed lies in the fact that the regulations impose that the FII have to be managed by financial agents without a real estate specialism required. The second is that the current operating process associates an FII with a specific asset, which limits the capacity of the instrument. Expert box 4.1 in this chapter provides a detailed view of the changes that are needed to improve the functionality of the FII.

Chile has also developed regulations for a REIT-like structure which is known as Fondo de Inversión Inmobiliaria, and has been operating in the country since 2007 (Fagenson, 2017). According to this author, Colombia also has a REIT type structure while Peru is looking at an appropriate legislation suitable for the countries. Expert box 4.2 presents a trust fund in Argentina for residential development, which, although not a REIT type, is a novel investment vehicle designed to combat inflation.

Expert box 4.1   Real estate investment trusts in Brazil

Claudio Tavares de Alencar

1. Structure and legislation

REITs in Brazil are joint-owned investment funds whose purpose is to apply funds in real estate undertakings and their derivatives. The majority of these Funds, called Fundo de Investimento Imobiliário (FII), follow international tendencies that focus on long-term income deriving from commercial real estate.

An FII is the typical structure available in the Brazilian market for investment sharing in real estate, which offers fiscal advantages not found in other forms of securitization in Brazil and whose structure differs from the REITs in the American market. These investment pools focus on the real estate market allocating resources to an investment manager. All the financial operations are regulated by the Rules for Fund Operation, where they are registered and submitted for approval by the CMV (*Comissão de Valores Mobiliários*/*Brazilian Securities Commission*), a government agency for regulating capital market operations, which in some ways is similar to the American SEC.

The legal concept of FIIs created some restrictions with respect to real estate maintenance, registration and transactions, as well as to certain business-development activities, which the FIIs are free to carry out. These restrictions were relaxed by Federal Law 8.668 (1993), which governs the creation and operation of the FIIs, and by CVM Instruction 205 (1994), which outlined the FII constitution, operations and management, providing management and operation norms in detail.

Law 8.668 (1993) defined that the operations (buying/selling of assets) and profit sharing of FIIs is tax-free, which was not the case earlier. Under current legislation, private investors are exempt as long as they comply with certain rules of distribution such as not owning more than 10% of the total shares in an FII.

Originally, FIIs' portfolios listed only real estate properties. Recently, however, portfolios may include a range of by-products such as participation in SPC real estate developers and even participation in receivables embedded in CRIs (Certificados de Recebíveis Imobiliários), which are the Brazilian MBS.

One other important difference between FIIs and the American REIT lies in the legislation covering management: [i] FIIs must necessarily be under the administration of financial institutions rather than commanded by an entrepreneurial type executive; [ii] the management of SPEs and REITs is in the hands of such individuals.

2. Management issues

The manager of an FII administers a portfolio having little manager's specialization, which, in real estate deals, should stand out with respect to specialised skills, not to financial management. FII administrators typically manage portfolios having little or no flexibility in that the FII is comprised of a single building. The administration of such a portfolio does not involve the dynamism of a continual change in portfolio positions, but merely requires topical actions and the administration of rental contracts and leases. For example, at BOVESPA [*The São Paulo Stock Exchange*] one can negotiate shares of an FII, the portfolio of which comprises a single office building leased to Caixa Econômica Federal (a state-owned bank) for ten years and with the tenant having the option to renew for another ten. In cases like this, the managerial skills required of the administrator comprehend collecting and distributing the monthly rental among the shareholders and renegotiating the price, perhaps every five years. The other FIIs listed at BOVESPA do not differ from such rigid patterns: they comprise office buildings (one per FII), shopping centres (one per FII), and one of them has a hotel in its portfolio.

When setting up a portfolio of commercial real estate, the REIT can be classified as either noble or poor.

A noble portfolio would consist of a group of properties belonging to a particular market segment (hotels, offices, shopping centres) so that any financial oscillations of any particular unit of the portfolio would be absorbed by the overall group and as such maintain a steady, balanced monthly income for the investor(s). A noble portfolio allows continual renewal of assets through buying and selling, whereas a poor portfolio only allows managers to sell a property when its market value has appreciated more than its capacity to generate income, which will in turn appreciate the portfolio's stock when such a sale results in the acquisition of asset(s) that are more profitable than those averaged by the group as a whole.

A portfolio designated as "poor" is that where the sole function of the REIT is to distribute the investment into one particular property, or in a small group of properties, therefore leaving the investor(s) vulnerable to oscillating market behavior, with no chance of hedging investments. Brazilian REITs can mostly be classified as poor, keeping in mind that the term "poor" does not in any way refer to the implicit quality of the portfolio's group of properties, but to the investment strategy used by the REIT. Those responsible for starting the REIT have chosen a primary method of creating opportunities where small and medium investors have a chance to invest in commercial real estate of significant financial expression, whose market performance (stable value and income) surpasses investing in small office properties, which are all that their small and medium savings may access. A global analysis of real estate investment strategies reveals that a share of an REIT with a non-diverse portfolio has a greater value than applications in small properties, although investing in an REIT having a portfolio with a diverse range of properties is, from the point of view of any conservative investor, much more advantageous.

An REIT shareholder receives regular dividends from investments in a way similar to those distributed by a Brazilian REIT (Fund), except the Brazilian model is a joint-owned investment fund (REIT) under the administration of a professional manager; whereas the model adopted by the US, Japan and European countries is a company that emits shares and is sometimes traded on the stock market, as is the stock of a Brazilian REIT.

3. FIIS and REITs size issues

American REITs, of which many are private, are the most relevant and experienced in real estate investment. In December 2015 there were 233 Trusts having capitalization of more than USD 938 billion. Forty-one owned mortgages (similar to the Brazilian REIT which has CRI) representing a 5.6% share of market. One hundred ninety-two owned buildings with capitalization of USD 886 billion. Here, 8% of the resources were invested in diversification and 92% applied in specific properties such as offices, shopping centres, hospitals, etc.

Examples: A regional headquarters building belonging to the state bank of Rio de Janeiro, Caixa. The bank deactivated the headquarters function in the building influenced by the REIT, which was under management of the bank itself, and the REIT then rented the building back to the Caixa Bank. Another example of this type of maneuver was the creation of an REIT with participation in Shopping Centre 13 in São Paulo. The REIT was specifically created to raise funds for completion of the project itself.

The average American REIT has a market value of USD 4.0 billion, whereas the FIIs on the Brazilian stock exchange are valued at R$196 million Brazilian Reais.[1] In relation to the Brazilian Stock Market volume of BRL 9.7 billion Reais, 129 FIIs have already been created, whereas the 175 American REITs on the Americans Exchange represent an Equity Market Capitalization (EMC) 62 times greater, resulting in an average 56 times greater.

There are some eloquent examples of the noble structures of American REITs; for instance, Simon Properties (shopping centres) alone has a portfolio 2.4 times greater than all the Gross Leasable Area (GLA) listed in Abrasce (Brazilian Association of Shopping Centres). Among the largest office holding REITs, Boston Properties, which is on the Stock Market and Private Equity Offices, has portfolios that value more than the entire area of Faria Lima Avenue in São Paulo.

The lack of attention developers have dedicated to the system of attracting funds by way of FIIs may be, in part, responsible for the slow evolution of the Brazilian market. If we consider the dimension and value of the REIT EMC, which is equivalent to 2.8% of the American GDP, this relation in Brazil would represent a market potential of R$60 billion, which is nearly 20 times greater than the market's current appraisal. Illustrating the magnitude of this potential is the region of Luiz Berrini Avenue, where the total value of new triple A office buildings does not exceed R$6 billion.

4. The futures of the FIIs

The question in Brazil is "who is responsible for reversing this tendency of creating small mechanisms for the sharing of investments that are sometimes only a part of a building or property, and take the necessary steps to establish a modern market where REITs represent authentic channels of investment in commercial properties?

Nobly structured REITs could promote the inclusion of commercial properties such as hangers, office buildings and even hospitals by taking advantage of the not too abundant but available resources that are sheltered in mediocre, high-risk investments such as the small offices that pop up and sell quickly in São Paulo.

The question is if chanelling available savings into modern, safer investment instruments is the role of the financial market or that of the real estate market. Modest efforts in this direction have been taken by administrators within the financial market (Real Estate agents and investment banks) to promote REITs with the proposition of creating investment portfolios of commercial buildings, which have an unexpressive volume of operations in relation to the size of the market itself. At the same time, however, the sale of stock in specialised property companies (BR-Properties, Iguatemi, BR-Malls, etc.) having the same objectives as an REIT but without the tax advantages and under greater management costs than an REIT, achieved a higher volume of trading than those engendered by the Funds that propose creating a portfolio of commercial buildings.

It is evident that brand specialization verified by administrators is lacking in Brazil; this in view of the capacity of even new brands (i.e. BR-Properties, BR-Malls) to seek large quantities of stock in investments in contrast to the inexpressive REITs focused on the creation of portfolios even in face of the tax advantages that incomes from REIT stock has in contrast to incomes generated by business associations. As long as administrators are focused on the launching of small commercial office space using traditional methods and do not understand that they could unite all the launchings under one REIT and offer the market an authentic product for investment and not just a common, everyday real estate investment, it seems unlikely that REITs will be able to escape the category of Poor Funds and migrate to a position in the market that is equivalent to the markets of first world economies.

5. FIIs profitability

We illustrate this section with a comparison of investments. One is a Brazilian REIT with a portfolio profile that generates income, which oscillates frequently because part of its equity is applied in a hotel that has seasonal proceeds. The other is applied directly or by way of the REIT in one individual office building. The two investments are compared against IBovespa (São Paulo's Stock Exchange index): i) in December 2004 the investor invests BRL1,000.00 (one-thousand Brazilian reais) in a rental property with annual income of 10% (0.797% nominal monthly equivalent); ii) the rent is adjusted once every 12 months according to the variation index of Ipca-ibge (Brazilian inflation index); iii) the investor receives the monthly rental income and pays 27.5% tax on this income; iv) the property appreciates in value in accordance with the curve of Ipca-ibge to the extent that, along with the income received during the six years, the investor winds up with an asset worth BRL 1,332.00 at the end of the cycle in December 2010; v) this same investor applies the same value (1,000.00 reais) in a share of a REIT on the same date (December 2004) and receives an equivalent monthly income; vi) as the administrative costs of the REIT are less than the 27.5% tax, the investor takes the surplus and buys more quotas of the REIT at market value, which is the original value adjusted by Ipca-ibge; vii) in this case the investor winds up at the end of the cycle (six years) with an equity of BRL 1,478.00; viii) this same application in a Brazilian REIT, whose portfolio has AAA offices rented out and a five-star hotel, produces income more erratic – when the net income is less than that offered by traditional property the investor sells shares at market value in this cycle and when there is an income surplus he buys shares; ix) in this case the investor winds up with an equity in the REIT of BRL1,870; x) for benchmark, for the Ibovespa application, with the investor buying or selling whenever there is surplus income or lack of it, and considering a 15% income tax, the equity would have reached BRL2,209.00 by the end of the cycle.

4.2 Countries implementing a "unit of savings" system

As mentioned in the introduction to this chapter, most countries battle with inflation and long-term real estate finance mechanisms. For instance the SFH applies a "referential rate" to maintain purchase power parity and has been doing it since 1964. Chile also introduced in 1967 of a system of savings accounts that are linked to inflation, the difference between this system and that of Brazil is that the latter has been operational since its implementation and is the one most South American economies are trying to emulate. The Chilean Unidad de Fomento (UF)[2] has as its main objective to keep real value of loans in volatile environments (Morandé and García, 2004). The UF is a financial unit of investment that is adjusted according to the consumer price index of Chile (IPC for its Spanish acronym). It is regulated by the Chilean Central Bank and is calculated using the following formula:

Daily adjustment UF = $(1 + pvIPC)$ 1/d
100 value UF in Chile

Where:

pvIPC = is the percentage value change of the IPC from the previous month
d = time period of the calculation in days

The result is the current value of the UF + the daily adjustment. Graphic 4.1 shows the monthly variance of the Chilean UF from April 2016 to April 2017 while Graphic 4.2 shows the monthly variance of the IPC.

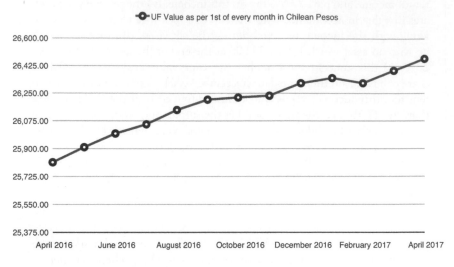

Graphic 4.1 Monthly variance – UF

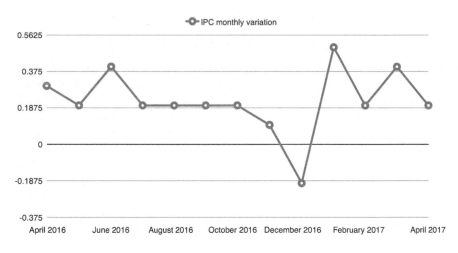

Graphic 4.2 Monthly variance – IPC

The original value of the UF was set at 100 Escudos (the official currency of Chile in the 1960s).[3] The UF has gone to receive quarterly adjustments when it first started, to be monthly and later daily adjusted to keep the real value more closely monitored.[4] The success of the UF in protecting savings from inflation and price fluctuations has meant that the unit is now widely used in Chile, not only for savings to acquire homes but also for current accounts and even to calculate monthly salaries of employees. In 2002 the government introduced other credit vehicles denominated in Chilean pesos with nominal interest rates and also credits with variable interest rates denominated in UF.

As seen in Chapter 3, the housing demand in Chile is serviced by the State and coordinated by the Ministerio de Vivienda y Urbanismo (MINVIU) and by the Servicio de Vivienda y Urbanismo (SERVIU). The different programs include: Viviendas progresivas and Viviendas básicas, both mainly targeting low income families. The State either acquires properties or commissions new development directly to the construction companies but in all cases the method for purchasing and developing is based on the UFs.

Other South American countries have followed the Chilean system, including Uruguay, Bolivia and most recently Argentina.

In Uruguay the system is known as Unidades Indexadas (UI-indexed units) and it is calculated by the country's Institute of National Statistics.[5] The UI was introduced in June 2002 and it has been gradually incorporated into the financial system of Uruguay. One of the latest regulations that has been implemented in support of the UI is the Law of Financial Inclusion (Ley 19,210). The law stipulates that the sale of goods and services in which at least one of the parties is a legal person, sole proprietor, de facto society, irregular company, civil society or similar, and where the total amount of the transaction is equal to or greater than 40,000 IU (about USD 123,000), the operation must be paid using electronic payment. In addition, in the sale of goods and services made by any natural or legal person, whose amount is equal to or greater than 160,000 IU (approximately USD 493,000), the payment can only be made through electronic means of payment (credit card, debit card, electronic money instruments and electronic transfers). The fine for not using electronic payment is 25% of the total amount of the transaction and is applicable to all parties involved in the operation. This affects all real estate transactions in Uruguay, including rents and sales. In the case of rents, the property owner is responsible for meeting the requirements set out by the new law. The non-compliance with the law in the case of property can carry other legal sanctions such as losing rights to eviction or application of a fine of three times the monthly rent. In the case of sales, operating outside the law might risk the validity of the property title in addition to the 25% fine mentioned above. The Law of Financial Inclusion is clearly a method of increasing transparency in the operations and reducing the informal economy. It should be accompanied with appropriate regulations to protect those at the bottom of the financial system who might struggle with transactions costs and taxation.

In Bolivia, the system is called Unidad de Fomento a la Vivienda (UFV) and is regulated by the Banco Central de Bolivia. In this country the unit is calculated on the basis of the Consumer Price Index (CPI) published by the National Institute of Statistics (INE). The UFV was created by Supreme Decree 26390 (November 8th 2001) and by resolution of the Board of Directors of the Central Bank of Bolivia (No. 116/2001 – November 20th 2001). In spite of its denomination, the UFV is not limited only to real estate operations. It serves as a reference for all types of financial transactions; for example, deposits and credits. The latest country to join the trend of indexed units for savings is Argentina, only that in here the unit is more closely connected to real estate.

4.3 The case of Argentina

From all the countries in South America, Argentina has one of the highest levels of inflation, ranking only second to Venezuela. For this reason, this section will look at how the country manages to fund long-term residential investment to support the housing policies in place since the appointment of the new government in December 2015.

According to data from the Subsecretaría de Vivienda y Habitat (SSV) or Ministry of Housing, the housing deficit of Argentina is of 3.8 million homes including 1.6 million in quantitative (new construction) and 2.2 as qualitative (in need of repair) deficit. To tackle this problem, the Government has an annual budget of AR$ 25 million to invest in social housing programmes for the whole country. All programmes available in the national housing plan target the middle- and low-income sector, which together account for 95% of the deficit. The middle segment of the population (35%) are households with an income between two and four minimum wages (MW) (AR$ 16,100 and AR$ 32,200) per month, while the lower sector (60% of households) have incomes of less than two MW (below AR$ 16,100).

Today most of Argentina's affordable housing programs are mainly accessible to the middle classes and the country is developing suitable programs to address housing needs for the most disadvantaged groups (i.e. those earning below 2 MW). The main financier of programmes is the State and delivery happens via the Institutos Provinciales, which are managed by the local governments in each of the 23 provinces of the country. In 2016, Argentina began looking to deliver via other channels that could potentially incorporate the private sector; simultaneously, the Government also looked at neighbouring countries to learn from their experiences in the financing and delivery of housing programmes.

As a result of these searches for new solutions, the Government introduced a savings system similar to the Unidaded de Fomento (units of savings) system pioneered by Chile and also in use in Uruguay and Bolivia.

4.3.1 Unidades de Vivienda (UVIs)

Households in Argentina are now able to save by means of Unidades de Vivienda (UVIs) – housing units. These are households' savings in pesos that have been "converted" into square metres of real estate. The value has been benchmarked following the construction costs of a pre-determined housing typology (known as "model 6" for a single family unit), which is measured by Argentina's National Statistics Agency (INDEC). The value is linked to the construction prices for the metropolitan area of Buenos Aires (ICC index for its Spanish acronym), which is also provided by the INDEC.

The UVI benchmark was established on the 31st March 2010 with an initial value of $14.05, and representing the cost of building a 1/1,000th of a square metre of housing. The value is obtained from calculating a simple average for the price of construction for the Autonomous City of Buenos Aires and the cities of Córdoba, Rosario, Salta, Paraná and Santa Fe, therefore contemplating a range of construction prices from the most expensive Buenos Aires to medium cities like Córdoba and less expensive locations such as Salta.

In order to adjust the UVI values to inflation, the Central Bank of Argentina (BCRA for its Spanish acronym) introduced on 31st March 2016 an inflation coefficient (Coeficiente de Estabilización de Referencia – CER) to be applied to the UVI. The new unit is called UVA (Unidad de Valor Adquisitivo) and provides the real interest rate of accounts as opposed to the UVI, which is nominal.

The two units (UVI and UVA) are now used for savings and to provide credits for housing. According to the BCRA, who publishes the values of both units on a daily basis, the current value of both units as per 22/11/ 2016 are:

> UVA Arg $14.05
> UVI Arg $ 17.16[6]

The UVI offers savings accounts linked to the ICC and with a minimum term of 90 days; while UVA offers savings accounts linked to the CER and a minimum term of 180 days. In both cases these are 0-fees accounts, designed specifically for holding either an UVI or UVA savings account. Commercial banks can also offer interest rates on top of the BCRA's base rate. Table 4.1 shows the changes in the value of the UVA and UVI since its implementation in April 2016 as published by the BCRA.

Table 4.1 Argentina's UVA and UVI values since implementation

Argentina's UVA and UVI values April 2016–June 2017

Month	UVA Value	UVI Value
April 2016	14,05	14,05
May 2016	14,41	14,41
June 2016	14,88	14,82
July 2016	15,39	16,17
August 2016	15,93	16,40
September 2016	16,36	16,65
October 2016	16,52	16,79
November 2016	16,63	16,95
December 2016	16,92	17,44
January 2017	17,26	18,12
February 2017	17,51	18,33
March 2017	17,68	18,48
April 2017	18,07	18,90
May 2017	18,48	19,09
June 2017	18,97	19,61

Source: BCRA www.bcra.gov.ar/PublicacionesEstadisticas/Principales_variables.asp

Since the establishment of these adjustment units in April 2016, Argentina's financial system has provided adjustable mortgage loans per UVI totalling AR$ 272.9 million and captured fixed-term deposits in UVI for AR$ 133 million. This is a completely new direction for the country, since indexation to any type of contracts was forbidden by law since 2002.[7] The restriction included any type of real estate development contract, as well as mortgages and rentals. The decree 146/2017 allows now for indexation of real estate mortgages for purchases and for new developments, although restrictions still apply to rental contracts.

A year since the implementation of the UVA and the modifications to the Pro-Cre-Ar plan, the banks have issued over AR$ 7,000 million in mortgages (Table 4.2). As reported by the local media, one of the largest commercial banks offering UVA mortgages for housing acquisition is Santander Río. Thus far, almost one in three UVA type credits has been issued by this bank. On average, the mortgages granted are for an approximate amount of AR$ 1 million, with a repayment period of around 19 years, and for an LTV ratio of approximately 58% of the mortgaged property. The average age of clients with a Santander UVA mortgage is 37 years, which shows the interest in a sector of the population that was unable to access any type of credit for the past 15 years as well as the potential for growth of this market, considering not only the space for catching up with demand but also new household formation. Table 4.2 above shows the names of the different banks offering UVA and Pro-Cre-Ar mortgages in Argentina, while the second column shows the number of mortgages granted. The last column provides the total amount of loans issued thus far.

The mortgage market in Argentina is still small if compared with other countries. According to country data from Global Property Guide, Argentina's mortgage market represents less than 2% of GDP while in Chile the percentage is ten times higher.[8] At local level, a debate has started among the private sector and different government arms, with some arguing that the UVA and UVI models are unfeasible, considering the levels of inflation currently prevailing in

Table 4.2 UVA mortgages in Argentina from April 2016 to April 2017

Bank name	Total mortgages granted in this period	Total value of loans (AR$ million)
Santander Río	1,395	1,867
Banco Galicia	505	593
Banco Provincia	2,003*	1,973
Banco Macro	580	651
Banco Ciudad	1,500	1,552
Banco Hipotecario	1,155*	860
Totals	7,138	7,496

Source: BCRA**

Notes:
* includes UVA and Pro-Cre-Ar mortgages
** as reported in local media www.lanacion.com.ar/2031183-los-creditos-con-uva-ya-suman-7600-millones

Argentina. Even those who agreed to the need for stabilisation coefficient to the UVI, pointed out that it was a mistake not to take into account the wage increase for indexation as well as inflation.[9] During the electoral campaign, the new government pledged to issue one million mortgages by the end of their term in office (2019). The new UVI and UVA system are now in place to help implement this target, but only time will tell if the programme can succeed amidst low wages, high inflation and constant increases in consumer prices.

4.3.2 Delivery and financing of the different housing programmes in Argentina

i) Pro-Cre-Ar

In order to activate the mortgage market, Argentina's Government has put in place national policies that aim at supporting the housing demand. As explained in Chapter 3, the Plan Pro-Cre-Ar was established during the Kirchner administration in 2012.[10] The plan survived the last government change in 2015, albeit not without modifications. For example, the new Subsecretaria de Vivienda y Habitat (SSV) or Ministry of Housing, is now the government arm in charge of the Plan whilst before it was under the jurisdiction of the Ministry of Finance. The scope of the programme was expanded as well, making it much more comprehensive and aiming to promote access to low- and middle-income sectors, and to improve living conditions in informal settlements across the country. As a result, the new plan subdivides the housing delivery in a four-part scheme, which includes: housing (concerning the units to be delivered); integral improvement of habitat (concerning the neighbourhood and including issues around connectivity); institutional strengthening (dedicated to boosting capacity at local and municipal level); and program management (dedicated to improving the monitoring and delivery of programmes at all levels of administration).[11] Finally, the funding of the programme, which previously included mostly national funds, now incorporates international financing via multilateral credit agencies such as the Inter-American Development Bank and the United Nations (UN-Habitat).

Credit available for Pro-Cre-Ar vary according to the scheme that it targets either qualitative or quantitative deficit (see Chapter 3). They include:

i) New homes: the maximum value of a home under Pro-Cre-Ar is AR$ 1,650,000 and the maximum subsidy provided by the government is AR$ 400,000. Beneficiaries must have savings for up to 10% or 15% of the value of the house to qualify for the scheme.
ii) Self-build: for a maximum value of AR$ 500,000 for the land; families can build a house of up to 80 sq. m in total. The maximum value of the house is estimated at AR$ 1,100,000. Interest rates given for this scheme are between 3% to 8%; cost of loan repayments must not exceed 25% of the wages of the mortgage applicant, and the maximum repayment period is 30 years. As with the New Homes scheme, the maximum subsidy provided by the government

is AR$ 400,000, although in this case the families do not need to have any savings in order to apply to a self-build loan.

iii) Home improvements: designed to help families to carry out necessary repairs and refurbishment as well as to connect houses to the national grid (water, gas and sewage). The beneficiaries must demonstrate needs (earnings below 3 MW), and the maximum amount of subsidy provided by the Government is AR$ 20,000.

iv) Serviced land: this scheme is delivered by the National government in partnership with the local governments. The local municipalities provide the land and the infrastructure for new homes. The applicant can then continue under the self-build scheme.

It could be said that the intention to expand access to housing found unexpected pitfalls, such as coordination difficulties between the official bodies involved in the matter, and the slow economic recovery required for the system to function properly. For example, the need to re-name the UVI aroused as a dispute over copyrights of the name UVI between a minister and the Director of the BCRA; new loan announcements by the Banco Nación, who offers an alternative product to the UVA and UVI accounts with a three-year, fixed rate mortgage which added to the initial confusion of offers. The National Social Security Administration (Anses) also joined in the announcements with the re-launch of the Pro-Cre-Ar Plan. In the midst of this debate, the National Congress managed to approve the plan for UVI savings and credits.

The lack of progress is particularly worrying in the interior of the country. In the province of Tucumán, where LINK operates (see Expert box 4.2), the Banco Nación has granted only 13 loans for purchases of new homes in 2016. According to local executives, this shortage is due to the increasing value of houses. The other problem identified is that applicants do not qualify for funding as monthly incomes are not sufficient to meet lender's requirements.

Expert box 4.2   The need for financing mechanisms for housing in Argentina: the case of the Province of Tucumán

Sebastian Piliponsky

According to the Argentine Government, there are 3.5 million families affected by the current housing deficit. The same source indicates that to solve the problem, 1.3 million new homes are needed, while a further 2.2 million informal dwellings are in need of upgrading and repairs (La Nación/"Es ley el plan de ahorro y crédito en UVI," 02.09.2016).

The housing deficit goes hand-in-hand with the idea that it takes about 18 years of work to acquire a small 50 sq m apartment in a prosperous neighbourhood, such as Palermo in the city of Buenos Aires. Indeed, the average price for these types of units are USD 140,000 (2016) while the monthly minimum wage is ARG$ 9,000 (equivalent to USD 566 in January 2017). This means that a family with two workers and a combined salary of two minimum wages earns USD 1,132, would find it extremely difficult to access the property market. The concept of paying a home in 18 years might sound like a good deal for international readers, but the difference in Argentina is that there is no financing and families must pay for the property up front if they want to buy in the open market.

The housing deficit problem in the open market is deepened by current high levels of inflation (40.5 % at April 2016 with an increasing forecasted curve) and hence increased costs of construction. With property prices rising in a context of stagnant wages and a depreciating currency, the outlook is bleak.

The departure from the currency convertibility regime (or parity between the ARG. Peso and the US Dollar) was in my view catastrophic. The economy crashed at the end of 2001, bringing down the entire financial system. The subsequent confiscation of deposits issued by banks to avoid massive withdrawals in banking accounts (similar to the freezing of assets experienced by Northern Rock customers in the UK in 2008), generated a deep mistrust in the banking sector, which remains to this date.

From 2003 onwards, the economy recovered by pumping up consumption. Still inflation soared and the informal economy increased to the point that experts estimate that two-fifths of the population is currently outside the formal economy now. In order to control currency devaluation, the administration of the then President Néstor Kirchner, imposed severe restrictions on the purchase of foreign currency (known in Argentina as "cepo cambiario"). This and other protectionists policies prompted the isolation of the country and blocked the development of investment alternatives for acquiring homes in the open market. Between 2002 and 2015, the only government-backed programme to provide housing solutions to disadvantaged families was Pro-Cre-Ar, as no other alternatives were feasible. Notwithstanding, Pro-Cre-Ar worked with the typical A+B+C system (Savings, Subsidy and Loan), which only works for the sector of the population that is in full employment and as stated above, two-fifths of the population were in the informal economy, therefore unable to have a proven record of employment and savings, which is one of the main requirements of Pro-Cre-Ar.

The housing deficit has become a significant social concern. The lack of progress is particularly worrying in the interior of the country, where for example in a province like Tucumán there is a housing deficit of 220,000 homes including new built and a stock that needs refurbishment (National Census, 2010 figures). In this city, the National Bank of Argentina has only approved 13 loans to buy houses in the year 2016. This low level of mortgages issued in Tucumán is due to the high prices of properties, but also due to the fact that most of the population are in the informal economy. It is in Tucumán where in 2010 I co-founded a real estate firm called LINK

Investments which, as we will see, aims at targeting the affordability problem so prevalent in the housing market.

Tucumán is one of the poorest provinces in Argentina in terms of annual per capita production. Here one-third of the population is in poverty. Therefore, developing and constructing real estate projects to sell houses in the open market, for those with starting household income of 3.6 times the minimum monthly wage, can be challenging. This scenario means working with extremely limited profit margins for the developer and competition forces us to reduce, reduce and reduce: space, costs, profits and even quality. Even so, the majority of the clients at LINK are investors, and in my experience investors in Tucumán care little about minimum standards; they are after a property that can offer the best rental income. In my view, this is like designing a pair of sneakers for someone who does not run. If the one who buys the shoes is not a runner, s/he will ask that the shoes be nice and pleasant to the eye, but they may not be comfortable.

From the point of view of real estate development, meeting the demand of investors in a trust fund is a problem because it generates an artificial customer, who is not the one who will end up using the home. It is worth pointing out that this type of demand for investment properties carries ever smaller margins for the developer and more unsatisfactory results.

This brings me to conclude that the affordable housing in Argentina has been distorted by the investors and deepens the problem to those who really need it. As the system works via trust funds (i.e. development projects only take-off the ground once a number of buyers have been secured); the longer the development period the better, as this brings more possibilities of finding suitable buyers. This scenario means that projects that can be normally carried out in 24 months take, in fact, a minimum of 40 months to complete. This allows developers to lower the quotas paid by potential buyers in order to sell units more quickly. However, long construction projects mean more expensive projects, particularly in a country with high inflation. As the land has been bought and construction prices go up, the only possibility to complete works is by reducing the quality of the product. Inevitably, the end-product is of a mediocre standard, as most units are purchased by investors (our non-runner buying running shoes). Their only interest is obtaining the maximum yield from the minimum amount of square metres.

It can be said that there is a distortion in which: a) properties are mostly sold to investors; b) construction times are extended and costs escalate; and c) the quality of the units is affected. In my view, the combination of all these factors distorts the market and in that distorted market we, the developers, have to work.

As a result of this phenomenon of what I call distortion of the Argentine real estate market due to the investors' appetite for returns, investment in construction in the country in the past 35 years has been the most profitable, surpassing other forms of more traditional investment such as commodities. According to the Edisur Group's Studies Department, each ARG

$1 invested in real estate in 2002, became in February 2014 ARG $16 (a return of 1,500%), proving a popular investment option against currency fluctuation and inflation.

Again this national trend is highlighted in Tucumán. Between 2006 and 2014, the average price of a house per square metre in the affluent Barrio Norte in the capital city of the province increased or maintained its value in comparison with the other regional agricultural alternatives: crops and fruits such as lemon, sugar cane, soybean. The value of the square metre of construction in constant pesos (that is to say, subtracting inflationary growth) has remained in a stable form since 2006, increasing to 20% in real terms. In Argentina, therefore, real estate assets have always proven to beat inflation. Among the other aforementioned investment alternatives in Tucumán, property has been the most stable and outperformed other commodities (soybeans grew in similar terms to construction, albeit with high fluctuations derived from international prices).

Notwithstanding, the rental income from properties is decreasing. Indeed, during the period 2004–2014, Tucumán went from achieving the highest rents in the region in comparison to five other locations (Jujuy, Santiago del Estero, Rosario, Mendoza and Santa Fe) to being the lowest. For investors, rent provides an annual income of 3% to 4% but the bulk of the capital increase is in the value of the property with a 30% annual growth. This can be explained by a new population trend, which saw that 14% of the population in Tucumán moved from living in houses to apartments in buildings; thus increasing the owner demand for blocks of flats, preferred by investors.

This scenario prompted us at LINK to develop the "Plan Puedo." The Plan functions as a modular funding system in a trust fund. The fund is intended to finance the building of three towers: Blue, Red and Green; with target completion dates of October 2020; February 2024 and June 2027, respectively. Each of the buildings is autonomous, but it is linked with the other two towers via the fund. The Plan consist of a deposit, which is currently the lowest in the market and can be paid over a year. This is followed by 120 monthly payments and a final redemption sum (key money) upon delivery. The monthly payments are 30% lower than others available in the open market.

For example, a studio apartment of 40 sq. m in the "Blue" tower in November of 2016 required an ARG $116,740 deposit; followed by 40 monthly payments of ARG $5,837. The buyer has then the option to move into the property by a payment of the key money (ARG $29,185 for the same example) and then pay the remaining 80 monthly payments of $9,631 each. As a comparison, similar schemes in the market require a down-payment of ARG $313,211 and 30 monthly instalments of ARG $15,661.

Prices are adjusted by following the price of construction per cubic metre of concrete. The price is adjusted monthly and published by the Argentine

Chamber of Construction; it includes two main variables: labour and cement cost. In the last ten years, the price of concrete increased at the same rate as inflation, so it provides a reliable method for indexation.

"Plan Puedo" has its own indexation method. For example, say an apartment sells for ARG $1,000,000 and a buyer has paid 40% of it, there would be an outstanding amount of $600,000. This outstanding is adjusted using the price of concrete per m3; for example if the price is at ARG $2,000, then:

> $600,000/2,000 = 300$ m^3 is the increment that will be passed on to the buyer by means of increased monthly payments. If the buyer still has 30 monthly instalments to pay for the property, then:

> $300/30 = 10$ is the additional m^3 of concrete that the buyer will pay on a monthly basis.

Until the delivery of the unit, the Plan works as a sort of savings program for the members of the trust fund. It is transferable and can be sold to a third party after due diligence and creditworthiness checks have been carried out on the new prospective trustee.

The trustee can also "move" to another of the towers within the scheme that are under development; or make advance payments to bring forward the delivery of the property. Monthly payments can be canceled by credit card or cash. In the event of sustained arrears or defaults in payments (due to incapacity to support the obligations assumed), LINK takes possession of the units and sells the property to a new owner. There is a penalty charge for the defaulter, but in no case is this greater than 20% of the value of the investment. The aim of this is to reduce litigation, relieve the burden on customers that become debtors, avoid passing on risks to the remaining trustees and ensure the completion of the project.

The greatest risk is the possibility of systematic default but this is considerably reduced if we take into account the importance of acquiring a home, a family will save, if what is at stake is the house itself. That commitment is enhanced over time, since having sustained the payment implies the development of a behaviour of savings and, on the other hand, the proximity of the reward (i.e. owning a property) reduces the default risk. But even if there is a 50% default in the project, a developer can successfully continue with the project by using the cash already generated by the project.

The first operations of the "Plan Puedo" provide a final illustration:

> A studio apartment of 41.53 square metres in the "Green" tower (6 ° B), with completion scheduled for 2027, was worth $582,001 in April 2016. In construction terms, this amount involved 348.5 m^3 of concrete. Those who entered the program under these conditions paid a deposit of $41,566 (equivalent to 24.89 cubic metres of concrete) with

120 installments of $4,158 and a "key installment" of $41,566. Ten months later, access to an apartment with similar characteristics (for example, 2 ° C of the "Green" tower) has an entry fee of $58,940, with 120 installments of $5,894 and a delivery of $117,880.

Only time will tell if "Plan Puedo" manages to facilitate access to a home in the open market to families in Tucumán at the lower end of the wealth pyramid. These are households who are currently either renting or living in overcrowded conditions with parents and extended family members.

4.3.3 Gradual move towards Public Private Partnerships (PPP)

Some governments, including Chile and Argentina, are interested in developing PPPs to expand the housing offer and reduce the deficit faster than if the sole provider was the Government. The routes for PPP in Argentina involve the following programmes: i) Prodevi, which aims at providing affordable housing with international funding; ii) Integrated Urban Developments (CUI for its Spanish acronym), which targets wealthier sectors of the population; iii) a scheme dedicated to trade unions' members; and iv) Pro-Cre-Ar. All programmes are regulated by law (Ley 27.358 and decree 118/2017). The decree regulates projects in the fields of infrastructure, housing, activities and services, productive investment, applied research and/or technological innovation.[12] The conditions for the contracts stipulate the need to partner with local talent, with minimum quotas of local labor and materials that must be met by the private partner.

i) Prodevi

Another government programme launched by the new Argentinean Government is the Housing Development Program (Prodevi for its Spanish acronym), whose implementation will be articulated with the current Pro-Cre-Ar Trust. The programme is aimed at families between two and four minimum wages (roughly with a family income of AR$ 16,100 to AR$ 32,200; see Chapter 1). The Government intends to agree contracts with providers during 2017, and deliver a total of 15,000 new homes, which will be phased in two stages: 7,000 units or more by the end of 2018 and another similar amount by the end of 2019. External funding is under discussion with China (Eximbank and Chinese Development Bank), and with it the Government wants to build 15,000 homes in three years – from the end of 2017 – beginning of 2018 – for the middle sector of the population where the housing deficit is of 600,000 units.

The average housing unit for Prodevi is between 55 and 60 square metres. The Government has announced that it will pay contractors up to USD 1,000 per square metre plus a contribution of 20% towards infrastructure cost. Mortgage conditions will follow Pro-Cre-Ar system and will be for a maximum loan of AR$ 1,650,000 and up to 90% LTV of the unit cost. The repayment period is up to 25 years, depending on an applicant's conditions, although the repayment/wages ratio will be kept to a minimum of 25%. The requirements and deadlines will be set by each participating bank.

ii) Integrated Urban Developments (CUI)

These are mixed-use, mixed-density, socially integrated developments located in infill sites in locations close to transport nodes and urban services. Private developers can provide the land or partner with the National Government to develop on public land managed through AABE (the National Agency of Public Properties). The aim mix includes:

> Social housing 20–40%
> Pro-Cre-Ar housing 20–40%
> Open market housing 20–40%

iii) Union members housing

This component provides housing for trade unions as well as armed forces and other civil servant groups (police force). The unions provide the land, municipal authorities pay for the infrastructure, and banks finance up to 70% of the total value of the construction via individual mortgages paid by the beneficiaries. To facilitate access to mortgages, the National Government funds 30% of the total value of the project via subsidies. Banks will approve and finance mortgages and also construction costs. Regional and city authorities are responsible for urban planning and infrastructure.

iv) Pro-Cre-Ar

This scheme also has been adapted to attract private investment. The Government offers fiscal land and up to 40% of the value of the project (VAT inclusive) to developers. In exchange the private developer agrees on a certain number of units to be given to the Government for families who qualify for the Pro-Cre-Ar plan. The criteria to select projects include: number of units to be delivered for Pro-Cre-Ar, solvency and track record of the development company and quality of the project.

4.4 Chapter conclusions

This chapter has presented the difficulties South American countries are experiencing when trying to arrange long-term financing for real estate in inflationary contexts. The move to expand towards PPPs will certainly relieve national states from having to finance the provision of housing. The experience presented by the private group LINK in Tucumán, Argentina (Expert box 4.2) shows the issues facing an economy with 40% inflation, where the real estate business is challenging not only for developers and investors but also for end users, who are the ones who will be eventually left with these poor quality assets. The sad reality of Grenfell Tower in London in the summer of 2017 shows the dangers of cutting costs in residential tower blocks. Events where new constructions have to be demolished because of poor quality materials in South America are not uncommon (Colombia and Brazil have some examples). The only way forward is to increase regulation and construction standards to guarantee that what is built today will last for generations.

Financial regulations are also needed for investments. As seen from the Expert box 4.1, the FII in Brazil can certainly go further if managers can specialise and if assets can be better classified. The next chapter (5) also presents experts' views that point out the need of regulation for investment in infrastructure projects, while Chapter 6 looks in more detail at the barriers investors face when trying to operate in the region.

Notes

1 The current exchange rate of 1 US$ = 3.3 Reais (R$) – source: Brazil Central Bank. Available at www.bcb.gov.br
2 The UF was established by Chile's Ministerio de Hacienda, Decree Number 40, 2 January 1967.
3 See Norm number 40 issued by the Government of Chile on 2nd January 1967: Artículo transitorio.- El valor inicial de la unidad de fomento será de E° 100 durante el trimestre calendario en que se publique este Reglamento en el "Diario Oficial." Available at www.leychile.cl/Navegar?idNorma=99246
4 See Decrees number 280, 12th May 1975 for the change to monthly adjustments and Decree 613 1st August 1975 for the daily one. Ministerio de Hacienda de Chile.
5 The UI was regulated by the Law 17.761 (12/05/2004). Available at www.ine.gub. uy/c/document_library/get_file?uuid=512842bc-4130-4883-860a-ea1e780f2ee2& groupId=10181
6 (www.bcra.gov.ar/PublicacionesEstadisticas/Principales_variables.asp).
7 See Articles 7 and 10 of Law No. 23,928, provided for in article 27 of Decree No. 905/02, ratified by Law No. 25,827, and article 21 of Law No. 27,271.
8 www.globalpropertyguide.com/Latin-America
9 As reported in real estate news agency Reporte Inmobiliario. Available at www.reporteinmobiliario.com/nuke/article3221-la-cuota-de-un-credito-ya-esta-muy-cerca-del-alquiler.html
10 Decree 902/12. Available at http://servicios.infoleg.gob.ar/infolegInternet/anexos/195000-199999/198531/norma.htm
11 See Decree 146/2017 which modifies Decree 902/2012 and in particular Expediente No. 17572/2016 cited in this document.
12 See Decree 118/2017. Available at www.boletinoficial.gob.ar/#!DetalleNorma/159233/20170220

Bibliography

Fagenson, Z. (2017) Eyeing the Office Market in São Paulo and Mexico City. Urban Land. *The Magazine of the Urban Land Institute*. Available at https://urbanland.uli.org/

economy-markets-trends/eyeing-office-market-sao-paulo-mexico-city/, accessed on 9th July 2017.

Morandé, F. and García, C. (2004) *Financiamiento de la vivienda en Chile*. Inter-American Development Bank. Available at https://core.ac.uk/download/pdf/6441339.pdf?repositoryId=153, accessed on 25th November 2016.

UQBAR. (2017) *Securizacão e Financiamiento Inmobiliario*. 10th Edition. Rio de Janerio, Brazil: UQBAR Publications.

5 Delivery and finance of infrastructure in South America

5.0 Introduction

This chapter outlines why the infrastructure needs to be delivered hand-in-hand with development, to ensure that the eventual development is sustainable, and also highlights how poverty and inadequate infrastructure are inexorably linked. It gives an overview of all types of infrastructure that is required to support real estate development. It highlights the need for innovation and new technologies that move away from traditional carbon-intensive solutions and looks at the level of infrastructure delivery across the Big Five in South America. The chapter then goes on to look at funding mechanisms such as government funding, private sector investment, public private partnerships arrangements and multilateral development banks loans. The chapter ends with an appraisal of different approaches to infrastructure development.

The quality of infrastructure is directly related to poverty alleviation. Poor transport and telecommunications accessibility reduces the access to opportunity and work that cities offer, and poor water and sanitation have negative impacts on health, which further aggravate the spatial poverty trap (Bird et al., 2010). Ultimately progress, both social and economic, depends on bringing people out of poverty and raising the overall prosperity of a country. Real estate development has a key role in partnering in the process of improving and delivering infrastructure. A study for the World Bank in 2011 identified a positive relationship between the level of infrastructure spending and economic growth (Calderón et al., 2011). The authors of the research suggest that doubling infrastructure capital raises GDP by 15%. The link between investment in infrastructure and poverty alleviation is now widely recognised by development agencies and governments and has been widely discussed (Bird et al., 2010). Expert boxes 5.1 and 5.2 highlight clearly the links while Expert box 5.3 points out the importance in understanding social needs and behaviour in terms of prioritising investment.

Planning and delivering infrastructure is, therefore, not only essential for development, but also for the economy. Infrastructure projects are long-term and they are often given the go-ahead for political reasons; it has not been unusual to see governments or mayors cancelling projects started by previous administrations.

For this reason, policy frameworks need to be agreed for the longer term (ideally 20 years or more) and the funding of critical infrastructure should be protected from the cycle of politics. Expert box 5.4 discusses the importance of policy in terms of delivery certainty.

When and how essential infrastructure is delivered and funded varies, depending on the political and policy environment that exists. The question of "who should fund infrastructure – the private or public sector?" is often related to the drivers of the development. If there is great social need, for example of affordable housing, then providing the infrastructure early by the public sector can act as a catalyst to bring forward private sector partners. However, if the market is driving investment opportunity, then the argument is most often that the private sector should pay. In the background there is always the experience that good infrastructure, particularly transport, will enhance land values; for this reason it is usually accepted that a developer should carry some or all of the cost of providing necessary infrastructure, but this has to be weighed against viability and needs to allow a developer to make a reasonable profit. (Harvey and Jowsey, 2004).

The Expert box by Nicolas Estupiñan (5.2) highlights that in Latin America economic growth has gone hand in hand with infrastructure investment, and that total investment in infrastructure has fallen since the late 1980s, but remained reasonably steady in the region from 1990–2005. The piece also draws attention to the fact that private sector investment in infrastructure in Latin America was significantly higher in the late 1990s (than the global average) and still remained relatively high into the first decade of the millennium. He stresses that the overall level of investment in infrastructure, whether public or privately financed, needs to be sufficient initially to deliver essential projects to tackle poverty and ensure sustainable growth, and in the longer-term to maintain and renew the existing infrastructure services.

5.1 Infrastructure situation, country by country

Table 5.1 shows that there is disparity in access to some infrastructure across South America. Whereas electricity and improved water quality is widely accessible in all countries, sanitation levels vary considerably with Bolivia having extremely poor access and Colombia and Peru only having moderate access. The transport and telecommunications sectors are generally poor; however, this is the perspective country-wide and cities will perform better. Notwithstanding, the depth and integration of transport varies considerably from city to city.

Not only do the infrastructure priorities vary from country to country, but Table 5.2 shows that the percentage of GDP that is spent on infrastructure also varies. It goes without saying that one would expect countries that have the greater needs for infrastructure are going to spend more than those that already have good infrastructure. If starting from a low infrastructure base, then it can be argued that a higher percentage spend is necessary. Also, when starting from a low base, the spending on infrastructure is likely to be a key element of poverty alleviation strategies and therefore a priority.

Table 5.1 Access to infrastructure by country

	Infrastructure: Access to water, sanitation, electricity and telecommunications					Infrastructure: Transport		
	Electricity	Sanitation	Improved water source	Telephone	Internet	Paved roads (%)	Rail lines km / 100 sq. km	Roads
Argentina	97	90	97	25	36	30	0.9	8
Bolivia	78	25	86	9	20	7	0.3	6
Brazil	98	80	97	22	41	6	0.4	21
Chile	99	96	96	20	45	20	0.7	11
Colombia	94	74	92	15	37		0.2	15
Ecuador	92	92	94	14	24	15	n/a	18
Paraguay	97	70	86	6	24	51	n/a	7
Peru	86	68	82	11	34	14	0.2	8
Uruguay	98	100	100	29	44		1.7	44
Venezuela	99	91	93	25	36	34	0	11

Note: Electricity data 2009; water and sanitation data 2008; telecom data 2010

Source: World Bank, *World Development Indicators* Database, December 2011

Table 5.2 Infrastructure investment as % of GDP by country

Spending on infrastructure (% of GDP), 2014	
Argentina	2.89
Bolivia	4.47
Brazil	4.1
Chile	2.83
Colombia	2.45
Paraguay	1.51
Peru	4.46
Uruguay	5.08
Regional average (% of GDP) (South + Central America)	3.4

Source: MercoPress 2014

5.1.1 Peru

The World Bank Overview, 2016, highlighted that Peru is the region's fastest growing economy, with poverty down, but it is suffering from slow execution of infrastructure, policy uncertainty and the impacts of climate related events. The 2015–16 World Economic Forum's Global Competitiveness Report,[1] ranked Peru 89th out of 140 countries for the quality of its infrastructure, below neighbours Colombia (84th), Mexico (54th) and Chile (45th). Peru does face an infrastructure gap, according to a study for Asociación para el Fomento de la Infraestructura Nacional (AFIN) by the School of Public Management at Universidad del

Pacífico. According to the study, Peru needs an investment of USD 160 billion to meet its needs by 2025; that would mean an investment of over 8% of GDP every year until 2025 (almost twice current spending).

Improving connectivity is a priority, particularly as it is felt that better road connections between medium sized cities would help to decentralise the country and promote development outside the capital city of Lima. However, due to the geography and topography of the Andes mountains and the Amazonian jungle, this is a major challenge. A key priority has been to revive the Costa Sierra road programme, a plan to connect some 27 roads in the Costa Sierra region to production centres in the valley (Miranda de Sousa Hernández-Mora, 2016). This project may have seen a setback, as Peru has been impacted by severe flooding attributed to El Niño. In early 2017, the region was hit by very high rainfall that caused extensive flooding and mudslides that killed over 100 people. The result was widespread damage to roads, railways, bridges and other infrastructure, with over 200 bridges and more than 2,000 km of highway having been wiped out. Minister Martin Vizcarra stated reconstruction was estimated at USD 6 billion, which is more than 3% of GDP. The disaster recovery is likely to take priority over previous plans for enhancing highway capacity.[2] In terms of rail infrastructure, ProInversión (see PPP 5.4.3), has plans to upgrade the 129 km rail line between the fast growing provincial high-altitude economic centres of Hyancayo and Huancavlica.[3]

5.1.2 Brazil

Brazil has prioritised investment in energy at the expense of public transport. It has the enviable situation now that 90% of electricity is from hydro-generation, but the lack of investment in transport means that there is poor public transport accessibility and the majority of trips rely on highways and air travel. As a result, highways are very congested, both with private cars and with freight trucks, with the consequence of poor air quality in many areas, and air travel is unaffordable for most low income people. Brazil has some 20,000 km of economically navigable waterways that are currently underused and the government is now recognising the potential of using the water transport to reduce congestion on highways by investing USD 6.2 billion to increase the use of waterways for transport and logistics from 13% to 29% by 2025.[4] Brazil also has 15 seaports or harbours. The telecommunications sector is well developed and according to government sources 45% of the population has access to the internet.[5]

The future challenge for Brazil is to develop well integrated public transport services that include mass transit and the development sector should be engaged in this process. Cities like Curritiba were at the forefront of introducing bus rapid transit (BRT) in the 1970s, which is a high-capacity bus service that operates independently of general traffic and more like a metro system. BRT is a very efficient way to deliver transport systems for cities needing quick solutions; however, they must be accompanied by a long-term vision that includes mass transit, as BRT ultimately has finite capacity.

Expert box 5.1 The contemporary infrastructure challenges facing Brazil

Professor Edmund Amann, 2017

After several boom years, the emergence of economic crisis in Brazil since 2011 has underscored some of the structural impediments to growth, which continue to plague Latin America's largest country. As Ferreira and Araújo (2006) make clear, there is a direct and very strong relationship between investment (or failure to invest) in infrastructure and economic growth. The growth elasticities here are especially pronounced in relation to investment in transport infrastructure (Amann et al., 2014). Yet, for a number of reasons that will be made clear, there has been a long-term failure to invest sufficiently in Brazilian infrastructure, whether in terms of addressing current demand or, perhaps more pertinently, towards raising longer term growth potential. Policy responses so far have failed to catalyse the necessary surge in investment; this is a situation that draws unfavourable comparisons with other emerging economies, notably China.

According to the well-known comparative indicator of infrastructural quality provided by the *World Economic Forum* in the Global Competitiveness Report Brazil's overall standing was at 120th place (WEF, 2015). The quality of its roads was ranked in 122nd place, its railroads at 95th, its ports at 122nd, its air transport at 113rd, and its electricity supply in 89th place. On a sector-by-sector basis, the scale and nature of the issues becomes more apparent still.

1) *Highways.* Although Brazil has the world's fourth-largest road network, there remain significant quality issues associated with it. According to a 2012 World Bank study, of 1.75 million km of highways, only 18% were paved.[6] This represents an especially significant deficiency bearing in mind that 60% of Brazil's freight moves by road. Moreover, in relation to optimal levels, the authorities have been spending far too little on highway maintenance. In the 2011–14 period less than 1% of GDP per annum was being set aside for such spending when, according to the World Bank, 6% of GDP would be necessary to catch up with advanced industrial countries. It is also important to note that the quality of highway infrastructure is heavily differentiated by region. While the South and South East are comparatively well served with divided multi-lane paved highways, the same is not true in some of the less developed regions of the country, notably the North, the North East and the Centre West. Even the capital, Brasília, remains to be connected to the South's multi-lane "interstate" system. Partly for this reason, transportation costs are notoriously high in Brazil: spending on logistics represents 15.4% of Brazil's GDP. In advanced countries this is typically closer to 8–10%.[7]

2) *Railways.* As is the case in most economies, the extent of the railway network is substantially smaller than that of paved roads in Brazil (five times smaller in fact).[8] If one adds the paved and non-paved road networks, then the rail network is no less than 50 times smaller. Unlike in other key emerging market economies – China and India for example – rail transportation is almost exclusively the preserve of freight, the latter being heavily dominated by iron ore (which accounts for no less than 79% of total rail cargo). Brazil possesses just 3.4 km of rail per 1000 square km compared with 14.7 km in the US.[9] While plans for a high speed line between Rio and São Paulo have been announced, financial constraints and delays in the planning process have impeded progress.

3) *Water and Sanitation.* According to Mourougane and Pisu (2011), "water and sanitation is the sector where investment is probably most needed" in Brazil.[10] In 2008, according to the World Bank Development Indicators, 80% of Brazil's population had access to 'improved sanitation facilities' compared with 96% in Chile (a regional leader), 83% in South America overall and 97.5% for the OECD countries as a whole. More strikingly only 47% of Brazil's population is provided with sewage collection, of which only 20% is treated.[11] There are also strong inter-regional variations in access to water and sanitation services.

4) *Airports.* The physical scale of Brazil, the absence of long distance rail services and the poor quality of highways infrastructure outside the South and South East mean that Brazil is highly reliant on air transportation. Here, as elsewhere, the infrastructure is associated with a legacy of under-investment and poor connectivity, placing Brazil at a disadvantage in terms of international trade, investment and tourism. Brazil's two most significant international airports, São Paulo's Guarulhos and Rio de Janeiro's Galeão, date respectively from the 1980s and 1970s and their terminals' capacity is now severely strained (Da Silva Campos and De Souza, 2011). At the same time none of Rio's or São Paulo's airports is currently served by rail or metro links: highly unusual for cities of such size and international standing. Size restrictions at these and other airports (notably São Paulo's Congonhas) together with limited capacity at the military-run national air traffic control system result in frequent delays and costs, which are passed on to passengers. In an attempt to overcome some of these issues, operating concessions have been granted to private sector consortia across a range of key airports (including Guarulhos and Galeão).

Source: Adapted from Amann et al., 2014.

The presence of infrastructural deficiencies on the scale depicted above is of prime national concern and, over the years, a number of policy initiatives have been launched to address the situation. One of the most high profile was the Growth Acceleration Programme, known by its Portuguese

Table 5.3 Brazil: PAC investments, percent by sector

Sectors	2007–10	After 2010	Total
Logistics	14.9	7.2	11.5
Energy	45.7	92.4	66.1
Social and Urban	39.5	0.4	22.2
Total	100.0	100.0	100.0

Source: PAC, Morgan Stanley LatAm Economics

acronym, PAC. The PAC was launched in 2007 at the start of former President Lula's second term. It continued into the administration of President Rousseff, who was impeached in August 2016. The PAC program in its latter stages prioritised investments in the following areas: My House, My Life (housing), Water and Light for All (water, sanitation and electricity), Bringing Citizenship to the Community (safety and social inclusion), Better Cities (urban infrastructure), Transportation (railroads, highways and airports) and Energy (renewable, oil and gas). Altogether, the planned spending under the PAC programs amounted to approximately USD 525 billion (2017 exchange rate). As well as providing for new infrastructure, the program was also designed to focus additional resources at operation and maintenance. These areas had been traditionally neglected in earlier infrastructure investment drives such as that of the early 1970s under the Second National Development Program (Baer, 2014).

The PAC proved reasonably successful in meeting its own targets. According to Amann et al. (2014, p. 15), "between the beginning of 2007 and the end of 2010, 82% of planned PAC 1 projects were completed with public investment rising to 3.2%[12] of GDP compared with around 2% prior to the program's launch," while "82.3% of PAC 2's projects had been completed by the end of 2013 with accumulated spending reaching R$773.4 billion, or 76.1% of the program's total budget."[13]

However, despite the undoubted successes achieved under the administrations of Lula and Rousseff on the infrastructure front, the PAC program did not prove a panacea; indeed, significant investment shortfalls remain in the urban transportation and sanitation sectors. Seeking to address this, the Centre Right administration of President Michel Temer (which assumed power in 2016) has attempted to kick start a program of private sector driven investment under the banner *Projeto Crescer* (Project Grow). The project is focused on the highways, airports and oil exploration sectors. It relies for its stimulus on a new round of privatizations and associated credit lines from the BNDES national development bank (Amann et al., 2017). Significantly, compared to its predecessors, *Projeto Crescer* places much greater emphasis on regulatory reform and the creation of a more predictable regulatory environment for investors. This speaks to one of the key factors responsible for holding Brazilian infrastructure investment back over the past few years. It is to this issue that the discussion now turns.

Long-term factors driving underinvestment in infrastructure

Regulation

Perhaps unsurprisingly, there exist a broad range of factors, besides access to finance, which account for Brazil's relatively unimpressive long-term track record on investment in infrastructure. A critical feature highlighted by many authors (e.g. Mourougane and Pisu, 2011 p. 17) is the high degree of uncertainty and delay surrounding the regulation and regulatory approval of infrastructure projects. This is especially true in relation to environmental licensing, a factor which has imposed severe delays on several power generation projects. The issue at play here is not the legitimacy of having an environmental licensing regime – or indeed any other form of regulatory regime. It is rather connected with the speed at which regulatory procedures can be completed and the predictability and transparency of the process. According to a 2008 World Bank study, at least 15–20% of the costs of Brazilian hydroelectric projects are accounted for by environmental licensing costs (World Bank, 2008).

Another important set of regulatory issues relates to the setting of tariffs. According to Amann et al. (2014), a key factor underpinning low rates of investment in highways and power generation in the Lula and Dilma years was the preference given to concession models in which the winners were selected on the basis of those able to offer the lowest user tariffs. Such a model did not prove conducive to attracting the required surge in investment. Neither did it provide sufficient basis for the adequate maintenance of critical infrastructure. In this connection it is perhaps no surprise that the regulatory models currently being developed by the Temer administration are noticeably more investor-friendly. Whether they will succeed in drawing in the required investment remains to be seen.

Finance

Brazil, as many other developing and emerging economies, remains savings constrained. Partly for this reason, and partly on account of a highly concentrated domestic banking system, the cost of raising capital via conventional private channels is very high by international standards. As a consequence, the provision of infrastructure – and large scale fixed investment generally – is largely reliant on just two financial modalities; funding secured internationally by foreign multinational corporations and, domestically, via the state owned national development bank (the BNDES) (Torres and Zeidan, 2016). While the BNDES remains a highly significant player in relation to infrastructure funding, its role is nonetheless being cut back by a new administration keen to rein in its influence and impact on the public sector balance sheet. Until the Brazilian domestic private sector capital market can be sufficiently broadened and deepened, infrastructural projects will remain reliant on an all too narrow – and vulnerable – channels of funding.

Technical capacity

Another key constraint impinging on the potential to raise infrastructure investment is lack of technical capacity. As evidenced by the presence of major home-grown construction multinationals such as Odebrecht and Camargo Corrêa, Brazil possesses world class engineering, architectural and civil engineering capability. However, partly thanks for a dearth of investment in training and education during the 1990s and early 2000s, such capacity is very thinly spread and presents real obstacles in the path of technically complex projects being rolled out on schedule.

Corruption

As is now widely known, the ramping up of infrastructure spending over the past decade has created opportunities for corruption and graft. Such opportunities have been seized by unscrupulous individuals, especially in relation to construction contracts and, most notoriously, in relation to infrastructure projects instigated by Petrobrás, the majority state-owned oil company. According to Amann et al. (2014), a study by the São Paulo-based FIESP industrial association estimated that with the money lost to corruption in the first stage of the PAC between 2007 and 2010, 124% more roads and 525% more railways could have been built (for details, see Funmi Ojo and Allison Everhardt, "Brazilian infrastructure and corruption," in *America's Business Intelligence*, Washington, D.C., October 11, 2013). On the brighter side, the rigor with which corruption allegations have been pursued by the Federal judiciary suggests that the incidence of corruption – and its associated costs – is likely to subside in the future.

Bibliography

Amann, E., Baer, W., Trebat, T., and Villa Lora, J. (2014) Infrastructure in Brazil's Development Process. *IRIBA Working Papers*, University of Manchester.

Amann, E., Baer, W., and Trebat, T. (2018 forthcoming) Infrastructure. In Amann, E. and Azzoni, C. (eds.), *Oxford Handbook of the Brazilian Economy*. New York: Oxford University Press.

Associação Nacional de Transporte Público (ANTP). (2012) *Sistema de Informações da Mobilidade Urbana – Relatório 2011*. Brasília: ANTP.

Baer, W. (2014) *The Brazilian Economy: Growth and Development*. 7th Edition. Boulder, CO: Lynne Rienner Publishers.

Baer, W. and Sirohi, R.A. (2015) Transportation Infrastructure and Economic Development: A Comparative Analysis of Brazil and India. *Global and Local Economic Review*, 19, 2, pp. 37–59.

Da Silva Campos, C.A. and De Souza, F. H. (2011) 'Aeroportos do Brasil: investimentos recentes, perspectivas e preocupações', *Nota Técnica No.5*, Brasilía: IPEA. Available at: http://www.ipea.gov.br/portal/images/stories/PDFs/nota_tecnica/110414_nt005_diset.pdf

Federal Government of Brazil. (2013) *PAC2: A gente faz um Brasil de oportunidades*, 6o Balanço, 2011–2014, Ano II. Brasilía: Federal Government of Brasil.

Ferreira, P.C. and Araújo, C.H.V. (2006) 'Growth and fiscal effects of infrastructure investment in Brazil, *FGV Ensaios Econômicos*, 613, pp. 1–31. Available at: http://bibliotecadigital.fgv.br/dspace/bitstream/handle/10438/483/2049.pdf?sequence=1

Ferreira, T.T. (2009) *Arranjos Institucionais e Investimento em Infra-Estrutura no Brasil.* São Paulo: USP, Dissertação apresentada ao Departamento de Economia para obtenção de mestre em Economia.

IPEA. (2010) Infraestrutura Econômica no Brasil: disgnósticos e perspectivas para 2015. *Livro* 6, 1, pp. 1–587.

IPEA. (2012–1) *Comunicado 128: A Nova Lei de Diretrizes da Política Nacional de Mobilidade Urbana.* Brasília: IPEA.

IPEA. (2012–2) *Sistema de Indicadores de Percepção Social – Mobilidade Urbana.* Brasília: IPEA.

IPEA. (2012–3) *Comunicado Indicadores de Mobilidade Urbana da Pesquisa Nacional por Amostra de Domicílios 2012.* Brasília: IPEA.

IPEA. (2012–4) *Transportes e Metrópoles: Um Manifesto pela Integração.* Brasília: IPEA.

IPEA. (2013–1) *Nota Técnica No. 2 – Tarificação e financiamento do transporte publico urbano.* Brasília: IPEA.

IPEA. (2013) *Territorio Metropolitano, Politicas Municipais.* Brasília: IPEA.

IPEA. (2013–2) *Nota Técnica No. 4 – Transporte Integrado Social: Uma Proposta para o Pacto da Mobilidade Urbana.* Brasília: IPEA.

IPEA. (2013–3) *Comunicado No. 161 – Indicadores da mobilidade urbana da PNAD 2012.* Brasília: IPEA.

IPEA. (2013–4) *Texto para Discussão No. 34 – Transportes e Mobilidade Urbana.* Brasília: IPEA.

Morgan Stanley. (May 5, 2010) *Brazil Infrastructure: Paving the Way.* Available at: https://www.morganstanley.com/views/perspectives/pavingtheway.pdf

Mourougane, A. and Piso, M. (2011) Promoting Infrastructure Development in Brazil. *OECD Economics Department Working Papers* No. 898, pp. 1–33. Paris: Organization for Economic Cooperation and Development.

Pompermayer, F. and da Silva Filho, E.B. (2016) Concessões no setor de infrasestrutura: propostas para um novo modelo de financiamento e compartilhamento de riscos. *IPEA Texto Para Discussão 2177 Fevereiro.* Available at: http://www.ipea.gov.br/portal/images/stories/PDFs/TDs/td_2177.pdf

Ros, J. (eds.). (2012) *The Oxford Handbook of Latin American Economics.* Oxford: Oxford University Press.

Serafim, M.C.S. (2009) *Análise das políticas para infraestrutura de transporte no Brasil a partir da década de 90.* Universidade de São Paulo, Escola Superior de Agricultura "Luiz de Queiroz," Piracicaba.

Spilki, M. (2012) *Public-Private Partnerships and the Role of Internal Control in the State of Rio Grande do Sul in Brazil.* The Institute of Brazilian Business & Public

Management Issues, The Minerva Program, The George Washington University, Spring 2012.

Summerhill, W.R. (1998) Railroads in Imperial Brazil, 1854–1889. In Coatsworth, J.H. and Taylor, A.M. (eds.), *Latin America and the World Economy Since 1800*. Cambridge, MA: Harvard University Press, pp. 383–405.

Tendler, J. (1968) *Electric Power in Brazil*. Cambridge, MA: Harvard University Press.

Torres, E. and Zeidan, R. (2016) 'The life cycles on National Development Banks: The experience of Brazil's BNDES', *Quarterly Review of Economics and Finance*, Vol. 62, November, pp. 97–104.

World Bank (2008) *Environmental licensing in Brazil: a Contribution to the Debate*. Washington DC: World Bank, pp. 1–37. Available at: http://documents.worldbank.org/curated/en/780411468236700081/Summary-report

World Bank. (2012) *How to Decrease Freight Logistics Costs in Brazil*. Washington DC: World Bank. Available at: http://documents.worldbank.org/curated/en/348951468230950149/pdf/468850ESW0P101000PUBLIC00TP390Final.pdf

Torres, E. and Zeidan, R. (2016) The Life Cycle of National Development Banks: The Experience of Brazil's BNDES. *Quarterly Review of Economics and Finance*, 62, November, pp. 97–104.

WEF. (2015) *World Competitiveness Report 2015*. Available at: http://reports.weforum.org/global-competitiveness-report-2015-2016/

5.1.3 Argentina

In Argentina, an energy and infrastructure deficit is a major development constraint. The World Economic Forum (2015–16) ranks Argentina 87th out of 140 countries for the quality of its infrastructure. Roads are congested, cargo train services are limited, summer power outages are frequent, and natural gas imports are at the maximum capacity that can be handled (Newbery, 2016). Table 5.1 shows that only 30% of the road network is paved. Growth in demand has outpaced domestic production for twelve years in a row. President Macri has announced plans to invest over USD 26 billion in infrastructure over the next four years and expects that the majority will come from the private sector. Projects include upgrading airports, freight rail lines and roads in the north, a new nuclear power plant, a gas line in the northeast, and several tunnels under the Andes for transport links with Chile. It is anticipated that infrastructure investment must reach at least 5% of GDP; this increase will be above the level of investment in recent years, but is less than the Argentina Chamber of Construction's estimate that spending needs to be up to 8% of GDP for the economy to grow at 5% a year. Despite the financial difficulties, Argentina has committed positively to move towards renewable energy and has set a target to get 20% from renewable sources by the end of 2025, mainly from wind generation in Patagonia.[14] However, as will be seen in the next chapter, the country should do more to meet its 2030 targets.

Expert box 5.2   Urban infrastructure in Argentina

Nicolás Estupiñan

Context

Developing strategic and sustainable infrastructure in Latin America has remained as one of the key challenges to address poverty reduction, improve productivity and competitiveness, and fight climate change. Several works analyse the so-called infrastructure gap in Latin America *vis a vis* developed economies, and the numbers continue to prove insufficient spending. Over the last decade Latin America has not been able to cross the 3% GDP barrier[15] and has not achieved the suggested 6.0%[16] to start closing this gap. One of the most relevant studies done on this subject was Calderón and Servén, 2010[17] where they show the positive relationship between the increase in the infrastructure stock and the economic growth in the region during 1991–1995 and 2001 and 2005. Standard and Poor's (2015) estimated the benefit over a three-year period (2015–2017) of an increase in spending on infrastructure spending of 1% of the GDP in the first year, finding multiplier effects on the economies of 2.5 in Brazil, 1.8 in Argentina and 1.3 in Mexico.[18]

As a result of this massive body of literature and research that evidences this strong relationship, countries in the region have undertaken fiscal efforts to increase infrastructure investments with mixed results. In the case of Argentina, despite its history of strong private participation during the 1990s due to the privatization processes, during the period 2002–2015 private investments fell dramatically. Table 5.4 shows how infrastructure spending, mainly public, has been growing over the last decade to reach levels closer to the 1990s' levels. Yet the 5% of the GDP goal continues to be unattainable.

CAF's commitment to sustainable infrastructure in Argentina

To overcome major sustainable development challenges, CAF[19] has focused its Country Strategy for 2016–2020 on two pillars: to foster Argentina's reinsertion on global markets as a high-value trade added partner, and to reduce social vulnerability, particularly in the north and in urban areas. Understanding the country's challenges, CAFs funding has been allocated to enhancing and fostering sustainable infrastructure mainly for water, transport and energy sectors. During 2016, two strategic operations were approved by CAF's Board addressing transport and water needs in the city and province of Buenos Aires. These operations are: i) the integrated management plan for the Lujan River Basin; and ii) upgrading the Belgrano Sur passenger rail.

Table 5.4 Infrastructure investments as % of the GDP, 2003–2015

	2003	2004	2005	2006	2007	2008	2009	2010	2011	2012	2013	2014	2015
Water & Sanitation													
USD million	55	89	126	221	331	504	566	648	963	1.267	1.452	1.498	1.293
% of GDP	0,04%	0,05%	0,06%	0,08%	0,10%	0,12%	0,15%	0,14%	0,17%	0,21%	0,23%	0,28%	0,23%
Transport													
USD million	325	618	1.092	1.713	2.133	2.709	2.627	3.266	3.941	3.601	4.332	3.561	3.400
% of GDP	0,21%	0,34%	0,49%	0,65%	0,65%	0,67%	0,69%	0,71%	0,70%	0,59%	0,70%	0,66%	0,61%
Total Infrastructure	0,64%	0,85%	1,42%	1,76%	1,50%	1,67%	1,82%	1,89%	1,99%	1,79%	2,23%	2,55%	2,31%

Source: Author calculations based on http://infralatam.info

Integrated management plan for the Lujan River Basin[20]

In 2016 CAF approved a USD 100 million loan to Argentina for management of the Lujan River Basin. This plan will prevent flooding, manage in a controlled manner the volumes of flow and moderate the effect of floods affecting ten counties of the province of Buenos Aires.

Over the last 50 years, the province of Buenos Aires (PBA) has been characterised by rainfall of great intensity. Similarly, changes in land use planning and the scant capacity of the drainage systems have influenced the overflow of water courses and the increase in the magnitude of the floods, with the subsequent flooding of urban and rural areas, as well as the closure of roads and routes. Flooding recorded in the Lujan River Basin has been especially significant in recent decades. This river rises in the confluence of the Durazno and Los Leones streams in the area around Suipacha and empties into the Parana de las Palmas River. It is a typical prairie river and possesses a sinuous channel of over 130 km long (80 miles), with a slow current and broad flood plains as a consequence of its scant fall. The basin is basically rural in its upper and mid sections, becoming more urbanised in the lower part of the basin.

The problem of flooding in the Lujan River Basin is linked to three major groups of factors:

(i)   Meteorological or climate ones, such as the increase in the volume and intensity of rainfall;

(ii) The physical characteristics of the basin as regards gradient, form and storage and flow capacity;

(iii) Human factors, such as changes in land use, growing urbanisation and the characteristics of the road and water infrastructure. In 2014, seven major floods occurred.

The Province has divided into two stages the implementation of the Lujan River Management Plan (LRMP), with an approximate total cost USD 313 million. The first stage, in the order of USD 158 million, proposes works and studies aimed principally at increasing the flow capacity of the river. The second stage, with an approximate cost of USD 155 million, and for which the province may eventually request new financing from the CAF, will complement the improvement works to the flow capacity, will include the construction of temporary overflow retention areas in the upper and mid parts of the basin, and the replacement of channel locks with inflatable dikes in Mercedes and Lujan.

Specific objectives of the Project are:

i)   Improve the flow capacity of the river, by means implementing complementary channels, and the enlargement and shaping of natural channels;

ii) improve current conditions of drainage in the basin, through replacement and expansion of bridges, crossroads and locks;

iii) propose land use and environmental management actions, in order to attain an efficient administration of the basin and to develop sound frameworks for land use management for water usage, environmental and socioeconomic needs; and

iv) implement an Early Warning System (EWS) that could alert population in case of an extreme climate event.

The direct beneficiaries total nearly 1.4 million people, belonging to the counties of Campana, Escobar, Exaltacion de la Cruz, Jose C. Paz, General Rodriguez, Malvinas Argentinas, Moreno, Pilar, San Fernando and Tigre, who live in the Project's area of influence in the mid and lower section of the basin. The indirect beneficiaries, considering the rest of the inhabitants of the listed counties and the population that lives in the upper part of the basin, are also estimated at 1.4 million inhabitants. In total, 2.8 million people, almost 17% of the total population of the province, will benefit from the plan.

This project has six components, but the main infrastructure works are in component 2: *Flow and channel enlargement works*. Enhancement of flow capacity of the Lujan River, through the implementation of complementary canals, the enlargement and shaping of natural channels, and the expansion of road crossings. At this stage the improvement of three priority sections are included:

i) The Santa Maria canal, which connects the Lujan River with the Parana de las Palmas River, and which will have an increase in its cross-section along its 7.1 km length;

ii) The channel of the Lujan River, from Provincial Route (PR) N° 6 to the bridge across National Route (NR) N° 8 in Pilar, which will be enlarged with trapezoidal sections for 12 km; and

iii) The channel of the Lujan River, from NR N° 8 to the railway bridge on the North Belgrano railway, that will be enlarged also with trapezoidal sections built along 9.5 km.

Table 5.5 shows the financial structure by component and funding source.

An economic evaluation for the project over a 20-year time horizon, identified the project's Net Present Value as USD 116.9 million, the internal rate of return as 20.2% and the cost-benefit ratio as 1.44.

Belgrano Sur passenger rail[21]

In 2016, CAF approved a USD 55 million loan to improve the mobility of the Belgrano Sur passenger rail (BSR) services and to enhance connectivity in the Metropolitan Area of Buenos Aires, Argentina. The project includes the elevation of 3.4 km of existing at-grade rail and the building of a 750 m rail connection to the Central Station of Constitución.

Table 5.5 Preliminary cost and financing of the project (US$)

Components	Sources (US$)		Total	
	CAF Financing	Local Contribution		
1. Engineering studies and others	1,665,000	1,665,000	3,330	2.1%
2. Flow and channel enlargement works	82,820,000	46,900,000	129,720	81.9%
3. Bridge replacement and enlargement works	11,040,000	6,210,000	17,250	10.9%
4. Environmental and land-use aspects	600,000	600,000	1,200	0.8%
5. Early Warning System	1,000,000	1,000,000	2,000	1.3%
6. Strengthening, supervision and auditing	2,000,000	2,000,000	4,000	2.5%
7. Financing costs	875,000	–	875	0.5%
Total	**100,000,000**	**58,375,000**	**158,375**	**100.0%**

Buenos Aires and its metropolitan area (RMBA) is one of the largest and densest cities in Latin America. Its population is over 15 million people, accounting for approximately 40% of the country, and covers an area of 200 km². Daily motorised trips are at around 24 million, of which 13.4 million account for individual transport, and 10.6 million for public transport. Its public transport system includes a bus network (340 lines), 831 km of urban railway, and 47 km of metro. From 1970 to 2000, the RMBA saw a doubling of numbers of private vehicles and the trend continues to grow. Two hundred fourteen million vehicles crossed the city access points in 2002, increasing dramatically to 394 million vehicles in 2013.[22] However, the public transport network has remained constant over the last 30 years. Since the late 1990s, investment, service, and reliability of mass transit have decreased, affecting particularly the railway system. This partially explains why the modal share for public transport went from 67% in 1970 to 40% in 2007.

Daily, the BSR carries approximately 10,000 passengers, mostly commuters accessing the city from the west part of the Province. For people from low- and low-middle income municipalities of La Matanza, Merlo and Moron, BSR is the best way to access the city. However, they experience a poor service, with low and unreliable frequencies, low speeds but at a very affordable tariff, around USD 0.30 cents. Currently, the BSR has four major level crossings and lacks a direct link to the Central Station of Constitución. It is estimated that over 336 user-hours per day are spent waiting at the level crossings.

Table 5.6 shows the financial structure by component and funding source.

Table 5.6 Preliminary cost and financing of the project (US$)

Components	CAF	Local	Total	% Share
1. Works	51,813,500	63,813,500	115,627,000	96.76%
2. Engineering studies and others	520,000	130,000	650,000	0.54%
3. Project management	2,149,000	556,500	2,705,500	2.26%
4. Other expenses	517,500	0	517,500	0.43%
Total	55,000,000	64,500,000	119,500,000	100.00%

An economic evaluation for the project over a 20-year time horizon quantified benefits including journey time savings for public transport passengers, reduced traffic fatalities and land value uplift in the surrounding urban areas. The project's Economic Rate of Return is 12.8% and the Cost-Benefit Ratio is 1.09.

The project, which will start implementation in 2017, is expected to improve mobility and accessibility of 1.5 million people living in the catchment of the BSR as well as reducing traffic fatalities and Green House Gas emissions.

5.1.4 Chile

Chile is currently well connected in terms of highways, railways, air and sea ports, and telecommunications. Ports are an important aspect of economic activity with 80% of foreign trade entering through its ports, and some are now developing passenger facilities to meet the demand of tourism and cruise ships.[23]

In 2014, President Bachelet launched Chile 30.30, a development and inclusion agenda that will invest almost USD 3 billion up to 2020.[24] Infrastructure investment is set to grow rapidly in the near future if major infrastructure projects such as the Costañera Central highway, the Santiago Metro line 7, the Maria Elena Thermo Solar Park, the Acueducto de Aquatacama, the Punilla Reservoir, the Arturo Benitez International Airport expansion and the Puerto de Gran Escala San Antonio development go ahead.[25] Also, 20 water projects including desalination and wastewater treatment projects are proposed for the north of the country.[26] Market research organisation, Timetrics, which has been tracking 350 major projects from announcement to execution, estimates that total value of investment in Chile is in the region of USD 140 billion. The power sector is the largest sector with over USD 91.4 billion, followed by planned water and sanitation projects worth USD 21.3 billion and roads projected to get USD 9.7 billion.[27] These infrastructure plans fall into three categories: government projects, PPPs and concession projects.

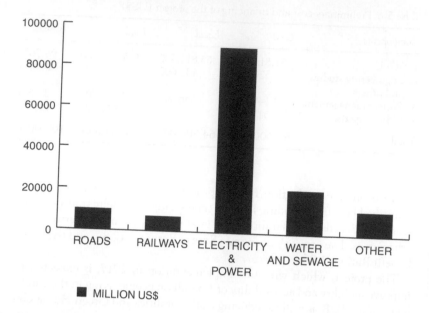

Graphic 5.1 Infrastructure construction pipeline in Chile, by sector
© Timetrics Nov 2016

For the capital city, the long-term transport plan, Santiago 2025, shows that roughly half the proposed expenditure will be in the city's Metro system, with 36% planned for "major" highways and a rather low 8% in total is proposed for bus, walking and cycling (see Expert box 5.3).

Expert box 5.3   Chile's twisting road to sustainable transport

Lake Sagaris, 2017

The search for scientific bases for confronting problems of social policy is bound to fail, because of the nature of these problems. They are "wicked" problems, whereas science has developed to deal with "tame" problems. Policy problems cannot be definitively described. Moreover, in a pluralistic society there is nothing like the undisputable public good; there is no objective definition of equity; policies that respond to social problems cannot be meaningfully correct or false; and it makes no sense to talk about "optimal solutions" to social problems unless severe qualifications are imposed first. Even worse, there are no "solutions" in the sense of definitive and objective answers.

Original formulation of wicked problems in planning
(Rittel and Webber, 1973)

From global institutions to cities of all sizes, finding ways to achieve "sustainability" at the local and regional scale is the quintessential "wicked" problem. The challenges of identifying and applying the tools, instruments and institutions required for the 21st century are deeply anchored in local-regional interactions. From this perspective, far from being "smaller" than the global, cities and their surrounding regions are strategic sites for experimentation and change, allowing multiple experiments in sustainability with lower levels of risk than attempts to impose untested, large-scale transformations from the top down.

Chile faces a particularly hard road in moving toward sustainability. The institutions inherited from the military regime (1973–1990) reinforced centralization, rigidification of public policy and reification of the so-called "free" market. Although most countries adopted some version of the powerful neo-liberal imagery[28] developed in Chile in the early 1980s, most saw important levels of contestation limit the model's application in significant ways, maintaining public health care and pensions in Canada and the UK, for example, or mechanisms for urban renewal, such as "eminent domain," and equity provisions in transport legislation in the US.

Governments in the post-World War II democracies of Canada, the US, Europe and Australia, moreover, retained significant powers over planning, enforcement and reassignment of goods and services considered important to the welfare of the general population and specific vulnerable groups, including human and civil rights.

In Chile, however, the combination of a brutal military regime at the height of its power and the neoliberal ideology propagated by Milton Friedman and the University of Chicago led to a refounding of the country and its institutions and laws (Délano and Traslaviña, 1989; Harvey, 2007). Severe human rights violations virtually destroyed a rich civic culture that had brought significant social gains from the 1950s up to the coup in 1973. This culture of active, capable, and committed citizens was forced underground during the regime, emerging in human rights groups, collective soup kitchens, protests and student battles, and other social movements in the 1980s (Oxhorn, 1995; Sagaris, 1996; Taylor, 1998; Taylor, 2004; Oxhorn, 2011).

A return to elected government in March 1990 improved many conditions, but involved no major reforms to an institutional design whose original purpose was to reinforce the former regime's power for an indefinite future (Huerta, 2000; Garreton Merino and Newman, 2001; Posner, 2003; Garretón Merino, 2009). Institutions, defined as "the formal rules, compliance procedures, and standard operating practices that structure the relationship between individuals in various units of the polity and economy" (Hall, cited in Sorensen, 2015) are the largely invisible but overwhelmingly powerful context in which this control is exercised.

As a result, the set of institutions governing transport planning and city regions such as Metropolitan Santiago (population 7.5 million people, 40% of GDP) became particularly fragmented and remain so. The Ministry of Transport and Telecommunications (MTT) is responsible for planning public transport (interpreted as buses and, in Santiago, Valparaíso, Concepción, also rail systems). Local municipalities and, more recently, the national Ministry of Housing and Urbanism, plan and execute individual cycle ways, usually without a master plan for creating an effective grid. The Ministry of Social Development controls a cost-benefit evaluations system that typically favours car-based transport projects, because on paper at least, these generate substantial time benefits (for high-income users). The Ministry of Public Works, particularly through its semi-autonomous Concessions department, plans and executes major infrastructure projects, which in the urban context are highways that favour a minority (only 40% of households in Santiago have cars).

Outside Metropolitan Santiago, regional governments consist of offices representing the presidency, with no planning powers, and without the democratic institutional arrangements that ensure representation and accountability to the region's citizens. This began to change in the 2013 general elections when, for the first time, regional councillors were elected by general suffrage. The regional council, or CORE, however, remains little known to most citizens, and continues to allocate significant funds primarily through trade-offs among political parties. Most urban planning in small and medium-sized cities and whole regions is done by regional secretariats of *each individual ministry*, with little or no coordination among them.

In the southern city of Temuco, for example, the regional housing ministry's office is implementing cycle ways on available tracts of land and some roads, often to vocal protest from the city government, and with little regard to public transport routes or users' actual needs or patterns of travel. Many are shared with pedestrians: despite the risks inherent in fast-moving cyclists (25–30 km/hr), many facilities mix cyclists with small children, baby carriages, and elderly pedestrians, making limited provisions for walkability or universal access for people with disabilities.

While Temuco's regional transport secretariat has developed two bus-only roads, with wider sidewalks and other improved facilities for pedestrians, and is contemplating a pedestrianization of the central area of the city, its counterpart in the public works ministry (MOP) is busy building a new bridge, underpasses, overpasses, and major highway facilities. These investments cater to cars and pay little or no attention to pedestrians, cyclists or the many users of animal transport (oxen, horses, and carts), despite the fact that in Temuco two-thirds of all travel is by walking (25%), cycling (2%), public transport (29%) and collective taxis (8%), and just one-third by car. Moreover, 35% of daily trips cover less than 5 km, distances more suited to active rather than motorised transport.

Trends

Figuring out how Chile is funding major transport investment poses challenges. A search for information reveals a colourful array of individual reports (www.mtt.gob.cl/vision-estrategica), covering the MTT's "strategic

Table 5.7 Estimated spending on transport by mode

Projects	US$ 650	Share of budget, %	Modal share %	Mode
Transport plans: Metro and trains	9,173,077	48%	6%	Metro
Concessions (major highways)	6,849,231	36%	26%	Cars
Roads (trunk and highways, 5%, plus local, 4%)	7,950,000	9%		
Bus corridors*	1,100,769	6%	26%	Buses
Cycleways	326,154	2%	4%	Bikes
Walking			34%	Walk
Total**	**19,161,538**	**100%**		

* As mentioned, the 2012 Origin-Destination survey combines Metro-bus trips into a single modal category making it difficult to identify the exact modal share for buses, bus-Metro, and Metro only trips, although the Informe Difusión reports that of the almost 26% of BIP trips, 52% are bus only, 22% Metro only, and 26% involve bus-Metro combinations. (p. 26, Informe Difusión O-D 2012, published in 2015).

** Difference due to rounding.

Source: Own elaboration using figures in a presentation of the Ministry of Transport and Telecommunications, Chile, Plan Maestro Transporte 2025, 2013

vision," its national transport policy, port and railway development, among other topics, followed by a second section with 13 documents covering road safety, enforcement, logistics, rules, laws, users and human resources.

Figures are seldom presented clearly, making it difficult to compare spending on specific modes with their modal share or number of users. Based on a 2013 presentation of the Santiago 2025 Transport Plan, however, this table provides some insight. Expenditure on the underground Metro and trains accounts for almost half (48%) of planned spending, despite the relatively low modal share. Metro's exact modal share is unclear, however, because the 2012 Origin-Destination (O-D) survey combines all fares paid by the BIP electronic card: in the previous O-D survey, Metro's share of daily trips stood at 4% (2006), with USD 1.5 billion in spending expected to double this to 8%. Comments from transport planners suggest it may not have met this goal, although during peak hours Santiago's Metro is now as overcrowded as Japan's, with six passengers per m^2 (Tirachini et al., 2013). In contrast, Santiago's cycle paths received just USD 48 million (2006–2010), a figure that doubled cycling's modal share from 2% to 4% by 2012. Informed estimates suggest it now stands at 6%. Public bike-share programs have enjoyed exponential success, to the point where memberships were frozen in late 2016, because the system couldn't absorb new users.

After trains, highway concessions (36%), trunk, non-concessioned highways, and local road improvements account for almost half of planned investments (that benefit mainly private cars – 9%). This car-centred approach has caused congestion to soar. In Santiago, half of car trips are less than

5–8 km, distances well served by cycling (cycle taxis, private bikes and bikeshare) or public transit.

Despite this focus of investment, powerful advocacy by a diverse array of pro-cycling groups, including women, has clearly pushed cycling up the political agenda. Despite considerable neglect, cycle ways that meet new national standards (2015) are emerging in most cities. Walking remains high, reflecting traditional urban supply patterns, based on neighbourhood groceries and local street fairs, which supply fresh produce and other goods in most neighbourhoods in Chilean cities.

The bad and the ugly

Probably the most disturbing elements undermining more sustainable urban policy is the Chilean government's inability to enforce its own regulations, particularly the ongoing erosion of its environmental impact assessment process and, more recently in proposed legislation, elimination of the EISTU, an assessment of the impact of new real estate projects on the urban transport system.

In many ways, Chile's urban legislation functions mostly as a statement of good intentions overrun by powerful developers fuelled by the huge funds generated by privatised pension funds and health insurance. In the Bellavista neighbourhood, for example, the regional housing secretariat ruled (three times) that the building permit for a six-story underground car park was illegal, on multiple grounds, particularly the lack of access roads with sufficient capacity, since the development is located in the heart of a walkable, human-scale heritage neighbourhood. But construction proceeds apace, despite the discovery of archeological artifacts, significant aquifers and other elements that would normally provoke serious concern.

To date, only one urban highway concession has undergone an environmental impact assessment (the Costanera Norte, in 1997): the rest simply pronounce themselves "express routes" and go through a simplified "declaration" procedure, rather than a complete study. This removes them from public scrutiny and the possibility of having to eliminate, reduce or mitigate the very serious environmental impacts of any major highway development.

The good

Sustainability, however, continues to gather force as an emerging issue. Citizen advocacy plays a growing role in demands for participation, transparency and policies more favourable to heritage, equity and environmental values. Cycling is the best example, as policies flipped from total neglect to national priority after a major, innovative participatory process brought together technical staff and citizen organizations, with support from Dutch experts (Interface for Cycling Expertise), from 2006 to 2010 (Sagaris, 2015).

By 2016–2017, transport planners in Santiago and Temuco were experimenting with similar innovations in participation for public transport. The inverted traffic pyramid (www.bicycleinnovationlab.dk/?show=jpn&l=UK), which puts walking and cycling at the top of priorities for transport space and spending, received widespread support among participants. Santiago centre, moreover, recently won a prestigious international award (from the Institute for Transport and Development Policy, 2016) for sustainable transport initiatives that have generated better conditions for buses, a more complete network of cycling infrastructure, and improvements to walkability and the public domain (see film: www.plataformaurbana.cl/archive/2016/12/19/video-la-transformacion-de-santiago/).

Coordination and collaboration to address wicked challenges

This overview of transport sustainability issues in Chile today reveals their complexity, particularly in a highly rigid institutional context. With no democratically accountable regional governments able to coordinate environmental, transport and land use planning, Chile is particularly challenged to come up with effective urban policies. This contrasts with institutions such as Transport for London, which is now able to coordinate citizen engagement with planning and implementation of projects that include public realm, pedestrian and cycle facilities, roads, buses, taxis, river transport, metro and suburban trains, accessibility issues, and even the transport museum.

What recent trends in Chile also illustrate, however, is that even in this difficult context, determined collaboration by visionary staff and politicians, working with active, well organized citizens can innovate against the odds, generating significant progress toward better urban living.

This, perhaps, is the greatest lesson from transport and planning in Chile today: while indeed urban sustainability is arguably our most wicked problem, human agency – people in different organisational niches working together for major change – can show the way.

Bibliography

Délano, M. and Traslaviña, H. (1989) *La Herencia de los Chicago Boys*. Santiago, Chile: Ornitorrinco.

Garretón Merino, M.A. (2009) Problemas heredados y nuevos problemas en la democracia chilena ¿Hacia un nuevo ciclo? *Sociedad y Profundización de la Democracia*. Instituto de Asuntos Públicos (INAP), Universidad de Chile.

Garreton Merino, M.A. and Newman, E. (2001) *Democracy in Latin America: (re)constructing Political Society*. Tokyo and New York: United Nations University Press.

Harvey, D. (2007) *A Brief History of Neoliberalism*. New York: Oxford University Press, 254 p.

Huerta, M.A. (2000) *Descentralización, municipio y participación ciudadana: Chile, Colombia y Guatemala*. Bogotá, CEJA.

Oxhorn, P. (1995) *Organizing Civil Society: The Popular Sectors and the Struggle for Democracy in Chile*. University Park, PA: Pennsylvania State University Press.

Oxhorn, P. (2011) *Sustaining Civil Society: Economic Change, Democracy, and the Social Construction of Citizenship in Latin America.* University Park, PA: Pennsylvania State University Press.

Posner, P.W. (2003) Local Democracy and Popular Participation: Chile and Brazil in Comparative Perspective. *Democratization*, 10, 3, pp. 39–67.

Rittel, H.W.J. and Webber, M.M. (1973) Dilemmas in a General Theory of Planning. *Policy Sciences*, 4, pp. 155–169.

Sagaris, L. (1996) *After the First Death: A Journey Through Chile, Time, Mind.* Toronto: Somerville House Publishing.

Sagaris, L. (2015) Lessons From 40 Years of Planning for Cycle-inclusion: Reflections From Santiago, Chile. *Natural Resources Forum*, 39, 1, pp. 64–81.

Sorensen, A. (2015) Taking Path Dependence Seriously: An Historical Institutionalist Research Agenda in Planning History. *Planning Perspectives*, 30, 1, pp. 17–38.

Taylor, L. (1998) *Citizenship, Participation, and Democracy: Changing Dynamics in Chile and Argentina.* Hampshire, NY: Macmillan.

Taylor, L. (2004) Client-ship and Citizenship in Latin America. *Bulletin of Latin American Research*, 23, 2, pp. 213–227.

Tirachini, A., Hensher, D.A., & Rose, J.M. (2013) Crowding in public transport systems: Effects on users, operation and implications for the estimation of demand. *Transportation Research Part A: Policy and Practice*, 53, pp. 36–52.

5.1.5 Colombia

Colombia was ranked 61st out of 140 countries for the quality of overall infrastructure in the World Economic Forum Global Competitiveness Report (2015–16). In the 2013 report, the quality of roads, railroad and port infrastructure, which were ranked as low as 130th, were found to be the most problematic. Colombia's geographical characteristics make it difficult and expensive to build transport infrastructure; as a result, it only has 900 km of railroads and trans-border transportation is limited.[29] However, public transport systems exist in the major cities; Medellín has a fully integrated multi-modal public transport system, including trams and cable cars. Bogotá introduced the Transmillenio bus BRT in 2000, but this network, which was highly successful in the early years, has over-reached its capacity and is now so oversubscribed that the city is having to develop a metro system. This is estimated to cost USD 13.2 billion, of which USD 9.2 billion will be provided by the Colombian government and the rest will be paid for by the district of Bogotá through multilateral bank loans (Johnson, 2017).

Expert box 5.4   Delivery of public transport systems in Colombia

German C. Lleras, 2017

It is common to see the national government leading large transportation projects in capital cities of developing countries; in Latin America, one recent example is the metro system in Lima. It is only at this level of government where the institutional and financial capacity typically resides. Moreover, these projects are part of the legacy of political leaders or parties. However, the real impact of these large projects is limited in coverage and creates new challenges emerging from the unclear allocation of responsibilities between national and local government. What is not that common is to see developing countries delivering a public policy whereby the national government works together with local ones to invest in public transport infrastructure and facilitate a transformation in the provision of affordable transit including, but not limited to, those very large projects. Colombia is one such example that differs from many others, perhaps with the exception of Mexico and Brazil; it is also an interesting case because such a program has already been in place for more than fifteen years and therefore it is possible to draw lessons and propose potential improvements.

The program finds its origins in the late 1980s and mid-1990s, when laws enacted by the Colombian congress allowed the national government to fund up to 70% of the debt needed to fund large transportation infrastructure projects. This was initially meant to fund the Medellín metro system, that started operations in 1995, and an eventual subway in Bogota. These laws defined technical, economic and institutional conditions to be met by local authorities for the national government to support the projects and make the corresponding appropriations. Notwithstanding this, the arrangement was only used for the Medellín subway; it was only after the implementation of Transmilenio, the ambitious BRT project in Bogota in 2000, that the country started to plan a policy using the existing legislation to support a national policy for the development of public transport systems across the country. This was finally written in a national government document issued in 2002[30] and was meant to create the framework for:

a) strengthening the institutional capacity to plan, manage, and regulate the privately delivered public transport operations;
b) create the incentives for cities to deliver an efficient public transport system;
c) break the cycle of providing only new car oriented road infrastructure by identifying other areas of investment;
d) create the incentives for the rational use of the available modes of transport including the private vehicle;
e) support locally identified initiatives that give priority to buses in the use of the existing road infrastructure;
f) develop the regulation needed to optimise the participation of the private sector; and
g) design the service thinking on the needs of the user.

By June 2015, when an evaluation of the scheme was made, eight medium sized cities and three large cities had signed agreements with the national government for funding of their public transport systems.[31] New local agencies were created leading the transformation of public transport, discussions with the private companies delivering the service started and for the first time, these cities had comprehensive analysis of their public transport, had proper designs for corridors and routes rather than following the requests made by atomised private companies that were not service driven but rather interested in increasing the size of the bus fleet, and finally had budgets to prepare a funding exercise to negotiate with the national government. In many cities construction and renovation of their main corridors was well in advance.

The process to implement the schemes has not been easy; in fact, it has faced many challenges ranging from real technical obstacles to others created by the change in local administrations, power struggles and the emergence of a fierce competition led by the motorcycle. A general evaluation of the scheme, made by the national government, was mixed and highlighted the following obstacles:[32]

- Lack of funding for relocation or construction of new utilities networks (power, water, gas);
- Time consuming interaction with utilities companies to solve issues arising from construction;
- Insufficient designs to contract works; and
- Lengthy processes to acquire or expropriate land.

In other instances, some cities could not gain access to national funding because, even when initially agreed, the local authorities did not allow for the matching funds to be made available. By mid-2015, ten cities had to review their funding scheme because approximately USD 200 million were not invested as planned. Following that analysis, and a refinancing of future payments, the program for these eleven cities up to 2021 is now made of USD 600 million.[33] This is a significant amount in cities that were used to invest less than USD 10 million per year in the same type on infrastructure. The evaluation was mainly made from the investment perspective rather than one looking at the results of the program; such an evaluation is still outstanding.

Thus far the programme has delivered in medium sized cities new or renovated road infrastructure, public spaces for pedestrians, bike paths and proper bus shelters and terminals. In some cases, it has also facilitated the delivery of new traffic control systems and a better relation with the public transport riders. In larger cities, it has spurred further development of BRT and cable car systems used for public transport; in this area Colombia is a leader with operating lines in Medellín, Cali and Manizales. More recently it has created the funding allocations for Bogota to continue the enlargement of the BRT network and the construction of the first line of its subway.

Of relevance is the development of cable car schemes. This innovative way to provide public transport is also a powerful way to make presence in hilly areas where the government has had a weak presence. Most of these areas started many years ago by informal occupation of lands and therefore did not have any planning permits or were thought as part of an integrated network of services and utilities. The challenge of transforming that type of urban fabric not only in its physical infrastructure is significant to municipal authorities. The larger network of such services is the one in Medellín, where four cable car lines are integrated with the subway providing reliable service to several neighbourhoods.

Having described in a general way the program it is important to acknowledge the main challenges that the Colombian cities face moving forward. The first two are related to the competing environment that public transport is encountering and is already reflected in declining patronage. The third is related to a missed opportunity that cities across the country are realising now and revolves around how to incorporate an integrated approach to develop public transport infrastructure and the general development of the land by working together with the private sector, becoming public developers or making use of existing value capturing tools. Finally, the fourth challenge resides in the institutional capacity of medium-sized cities and how this is affecting and will affect the further development of the program.

Colombian legislation allows for subsidies from the national and local government funding investment in infrastructure; the national government is not allowed to subsidise the operation costs of the service in any city while local governments are allowed, but do not do so. The national government guidelines for setting up the value of the fare define a cost structure and a negotiation process with the private companies delivering the system that does not happen. Thus, public transport fares do not accurately reflect the cost of provision and do not consider affordability as one of its premises. Efforts to create local allocation of funding to subsidise operation have been made in many cities but have not been successful because of the lack of a transparent framework to use them in the more efficient way. I argue that unless public transport quality is defined in a way that is partially detached from the full recovery of costs through the fare collection, as it has been thus far, the service will be behind competing modes. In medium sized cities where congestion is still not an issue, the "battle" against the car users is almost lost and the public transport rider that was considered captive now has other alternatives. The main one is the motorcycle; this mode offers two alternatives. The first one as a private mode paid by many in instalments cheaper than two daily bus ride fares; the second one is the motorcycle taxi that, even though it is more expensive than the bus fare, it provides door to door service. This last fact becomes paramount given the lack of proper road infrastructure, the hot and humid climate and in many instances the street violence that persists in the periphery of many cities. For the local and national government investment to be socially profitable, a new scheme to define the fare and opening the door for controlled operational subsidies is required. The incentives must be in place for the city to function in a way that farebox recovery is high, but this should be a policy objective rather than as an input to the process.

The second challenge is directly related to the one above: how cities can be built in a way that public transport is the most obvious and preferable option for the daily commute. Integration of investment in infrastructure must be linked to the development of other public buildings providing services and moreover to housing, commercial and employment opportunities. Large cities like Bogota and Medellín know how painful it is to make very long commutes in stressful conditions and are now trying to reverse that way. Medium-sized cities must think of a better way to prepare the city to be occupied by integrating land development regulation and location of higher densities around those corridors where investment is being made. There is no clear relation between the role of the government and that of the private developers to build a compact, transit oriented city. That is lacking in our effort to produce better liveable cities and may jeopardise the economic and social returns of the investments and of the policy itself. Colombia has had a history of creating different land value capture tools; in fact, many cities have used them in the past to fund infrastructure investment; however, it has been increasingly difficult to use those that resemble tax schemes and new ways of sharing the benefits of those investments, that are now mainly in the hands of private developers, are needed.

The final challenge is one of institutional nature and related to the difficulty in creating sound agencies that can deliver the programmes while strengthen the capacity to provide a sustainable service. Medium-sized cities compete in the labour market with the big cities at the national levels. The best universities offering the type of education needed to support the public transport program are not in the medium-sized cities and salaries and opportunities for professional growth are in cities like Bogota and Medellín. The national and local governments must also pay attention to this component because ultimately the money will only be well invested if the people leading the delivery of the program, contracting, building and designing further phases are located on the ground in each city.

As a conclusion, it is fair to say that Colombia has created a policy and delivered successfully some parts of the components needed to contribute to better cities through the development of public transport systems. However, the jury is still out in terms of the real success of the impact that this will have in making cities more liveable and sustainable. Medium size cities can become more attractive if steps towards these goals are taken and large cities are less successful in dealing with their own challenges.

5.1.6 International projects

A cross-border, interoceanic, trans-continental 2,300-mile-long railway to connect Peru's Pacific coast to the Brazilian Atlantic cost, to improve logistics between the north of the continent and the rest of the world, is under consideration. However, there are widespread environmental and social concerns regarding its impact. An initial memorandum of understanding for its development between Peru, Brazil and China was signed in 2015[34] but failed due to costs and impact concerns. A new agreement has now been signed between Bolivia and Germany for technical assistance and financing of the project (Ramos, 2017).

5.2 Main issues and good practice shaping infrastructure decision-making

Major infrastructure projects regularly take ten to 15 years to develop and deliver, but real estate development may only take half that time, so getting infrastructure delivery in sync with real estate development comes down to coordination of policy and planning. This need for synchronisation of delivery places huge challenges on political decision-makers and funding agencies. Ultimately the delivery time and possible delay of exploring innovative solutions needs to be weighed against longer-term sustainability benefits. UN Habitat (2011) estimates that 70% of all greenhouse gas emissions are attributable to cities. It is generally now acknowledged that CO_2 emissions must be reduced as they are the main cause of climate change (see Chapter 7). Tackling this issue is inextricably linked to how new development is planned. By promoting innovation in infrastructure delivery, new practices and technologies will emerge that minimise CO_2 emissions.

However, transport is a significant contributor to greenhouse gas emissions and pollution. It is responsible for 23% of CO_2 emissions globally and 30% in the OECD[35] and it is expected to grow to 40% by 2030 (ITF, 2010). The road sector, including freight haulage, is the biggest contributor to emissions and harmful particulates to the Earth's atmosphere. New development will generate traffic, but every development presents an opportunity to propose cleaner transport and more efficient solutions for freight and waste, as well as reduction strategies. In the coming decades, as cities grow, highways will see increased demand as more service, delivery, security and emergency vehicles are needed to support the growing populations. Highways will need to see a reduction in personal transportation vehicles and drastic changes in mobility will be needed to control congestion, which has the added problem of generating further pollution and negatively impacting air quality. It is important to comprehend the consequences of not taking action; carrying on "business as usual" will not be an option. New mobility solutions that do not rely on car-based travel will have to be part and parcel of development planning.

5.2.1 Mechanisms for delivery

Early clarity around the question of essential infrastructure and its funding is crucial. This way optimum solutions can be developed and opportunities exploited to the full. Once the level of infrastructure that is required, and its phasing, has been established, it is possible for the public and private sectors to engage and agree on potential partnership arrangements for the construction and the funding (i.e. privately, publicly or as a joint venture).

5.2.2 Holistic approach

The various infrastructure systems that make a development work are in many ways connected, so their design should be conceived holistically. This requires strategic planning so that systems can achieve maximum efficiency. For example, conventional sewage treatment is energy demanding, but by switching to anaerobic digestion, the process can also become a source of energy. Anaerobic digestion is a process for bacterial processing of sewage, which produces biogas that can be used directly as fuel in combined heat and gas engines and the digestate can be used as fertiliser.

5.2.3 Urgency: expediency or innovation?

In order to deliver infrastructure quickly there is inevitably a temptation to create sub-standard projects that rely on tried and tested technologies such as coal fired electricity generation, which are now more and more coming under scrutiny, when viewed against climate change mitigation targets (Stern, 2006). There is a tension between improving traditional solutions and finding alternative solutions that use clean energy and work with nature. This is mainly because many of the emerging green alternatives have not yet stood the test of time and are often initially more expensive, and there is always pressure to opt for the old and proven systems. However, the cost of renewable energy is coming down and according to WEF (2016), solar and wind energy is now comparable or cheaper than new fossil fuel capacity in more than 30 countries. In 2015, Uruguay had already transitioned to 94.5% of its electricity coming from renewable sources, which include hydro, wind, biomass and solar; and Paraguay stood at 90% renewable from a single hydropower dam.[36]

5.2.4 Funding and innovation models

Funding will always be the main hurdle to getting projects off the ground and solutions for securing monies need to be innovative. Engaging local communities as innovation and delivery agents can potentially reduce costs, but this does require new approaches to project management. Risk averse banks and funding agencies need to recognise the potential resourcefulness of local knowledge, rather than seeing community engagement as an obstacle to smooth delivery. Another approach to finance public works is through selling municipal bonds. In Argentina many provinces have recently sold bonds to finance public works, and municipalities are expected to follow suit. Though bond issues are more expensive than financing from tax collections, mainly because of the added cost of debt servicing. However, the advantage on leveraging projects is that the cities receive all the funds upfront, making project planning more secure. A concern, however, in Argentina is that the debt-to-GDP ratio is increasing as the economy is not growing as fast as hoped, given the high interest rates and the sharp devaluation of the currency at the end of 2015 (see Chapter 1). As shown in Table 1.22, ratings agencies give this country a positive outlook. However, construction activity dropped 12% in the first half of 2016, making construction delivery difficult (Newbery, 2016).

In 2015, in Peru, the Nuevo Metro de Lima consortium (delivering the new Line 2), raised USD 1.16 billion, with a 19-year bond offering a 5.875% yield, to help support the cost of the project: the highest non-governmental bond offer in Peru at the time (Miranda de Sousa Hernández-Mora, 2016).

Another mechanism for project funding is to capture the increase in land values that will certainly arise as a result of investment in improved infrastructure (see Chapter 4 for financing mechanisms of this type applied in Brazil).

Table 5.8 shows that across the region the level of investment as a percentage of GDP (for years 1981–2006) ranges significantly from 2.1% in Peru to 5.1% in Brazil, but since 2016 Peru is expected to be investing 5% of GDP on infrastructure with the regional average being around 3.5% of GDP (Newbery, 2016). Argentina, if it were to commit to all its infrastructure projects, would be spending 8% of GDP, which some observers deem is necessary (see Section 5.4.3, PPP). A close look reveals that Brazil invested heavily in electric power, but saw investment in other sectors decline. Overall, across the region, the priority investment has been in the power sector, whereas transport and water and sanitation have seen comparatively low investment (see Figure 5.1). However, in Argentina, 2003–2015 has seen a positive trend of rising investment in water and sanitation and transportation from 0.64% to 2.31% of GDP.

Table 5.8 Investment in infrastructure in Latin America, 1981–2006

Investment in infrastructure in Latin America, 1981–2006 (% of GDP)

Country	Period	Power generation	Land transportation	Telecommuni- cations	Water & sanitation	Total infrastructure
Argentina	1981–6	1.53%	0.81%	0.30%	0.12%	2.78%
	2001–6	0.50%	0.68%	0.38%	0.10%	1.67%
	Change	–1.02%	–0.13%	0.08%	–0.02%	–1.09%
Brazil	1981–6	3.30%	0.86%	0.74%	0.30%	5.15%
	2001–6	0.63%	0.41%	0.78%	0.28%	2.11%
	Change	–2.67%	–0.41%	0.06%	–0.02%	–3.03%
Chile	1981–6	1.65%	1.04%	0.47%	0.29%	3.44%
	2001–6	1.84%	1.69%	0.90%	0.78%	5.21%
	Change	0.19%	0.65%	0.43%	0.49%	1.77%
Colombia	1981–6	1.56%	0.94%	0.32%	0.31%	3.13%
	2001 6	0.58%	0.67%	1.01%	0.50%	2.77%
	Change	–0.98%	–0.27%	0.69%	0.19%	–0.37%
Peru	1981–6	1.35%	0.36%	0.32%	0.08%	2.11%
	2001–6	0.44%	0.37%	0.64%	0.04%	1.49%
	Change	–0.92%	0.01%	0.32%	–0.04%	–0.62%
Weighted Avg (By GDP)	Change	–1.40%	–0.52%	0.20%	0.00%	–1.73%

Note: Figures are total public and private investment

Source: World Bank, infrastructure in Latin America: an update 1980–2006, Cesar Alderon & Luis Serven, 2009

5.3 Considerations by sector

5.3.1 *Transport*

As mentioned at the beginning of this chapter, transport infrastructure is directly related to a city's economic performance and tackling poverty. A key issue, particularly for low-income workers, is the cost of urban transport; if the cost is unsustainably high compared to wages, it will fail to provide a route out of poverty for the urban poor. Travel times to work are often very long due to poor transport infrastructure and lack of land use planning controls that have allowed urban sprawl. As is highlighted in Chapter 1 and Table 1.18, it is clear that for many on minimum wages the cost of transport is a significant budget burden, with Brazil, Chile and Colombia being the most expensive in the region. The reason why, on the whole, is that public transport is heavily subsidised in South America.

Public transport in South America has a predominance of buses, often with poor performance (due to traffic congestion) and lack of capacity. Most cities lack urban or metro rail. In recent years many cities have seen a growth in informal transport systems, which further exacerbate traffic congestion, which has a high environmental cost and impact on air quality, which in turn affects physical and mental health. These services are operated separately from the formal services and offer "on demand" services mainly by minivan. There has also been a significant rise in motorcycle use with resultant high road traffic accidents (traffic death in Argentina, Brazil and Chile have been graded with the maximum level of risk by some agencies (see Table 7.1 in the next chapter for more details).

Improving public transport is a major challenge in many cities in South America, the key issue being lack of funds for maintaining and rehabilitating existing services or for investment in new systems, and there is heavy competition for what little funding exists. Another issue that frustrates delivery in many countries is the multiple layers of governmental control from the centre, the state and municipal levels and the lack of metropolitan transport authorities coordinating and optimising transport services. For example, in Colombia, rail is in the control of the state, but buses are controlled locally. Probably the most challenging issue is the lack of integration and alignment of land use planning, public transport planning and air quality control, as well as the lack of integration of systems themselves, including ticketing policy. Furthermore, there is very scant data available on travel patterns, so understanding the priorities and the big picture is difficult.

Poor investment in transport services and rising fares has resulted in a vicious circle emerging in many cities, where the car becomes the favoured means of transport for those able to afford it, thus driving up congestion, which in turn slows down bus services and makes them less attractive. This leads to less demand, decreased revenues and higher fares, rendering the services unaffordable and resulting in social exclusion and poverty. Low-income groups complain of lack of accessibility, affordability, availability and acceptability of the transport services, often only being able to afford the fare one way and having to walk home. In many cities 40% of people are walking to work.[37] The emergence of integrated ticketing systems is beginning to tackle this issue.

If development is not planned alongside public transport, ever increasing volumes of car-oriented developments will be created, which will further exacerbate congestion levels with associated problems. Transport planning, delivery and funding is inevitably very complex and adds financial and programming risk to projects, but retrofitting transport infrastructure after a development has been built is even more complex, sometimes physically difficult or impossible and certainly more expensive than building it before or during the construction of the development. By having a long-term framework for the future transport needs, it is possible to phase in public transport solutions over time, starting with quicker and cheaper solutions, such as bus services, but with a commitment to transition to fixed high-capacity solutions when a development is complete and fully occupied. This does require clear safeguarding of land by the state for the implementation of future transport infrastructure.

The issue of subsidy for public transport services is a fundamental question. Subsidies can either be targeted at specific services or users, or they can be general 'blanket' subsidies (as is common in Europe). Some argue for no subsidy, but very few public transport services are able pay for themselves, though there are exceptions (see below), and without the subsidy the fares would be unaffordable. Ultimately there has to be a balance between financial sustainability and the need to provide affordable services, particularly in low-income areas. Subsidies for transport are used widely across South America, primarily as poverty alleviation mechanisms (see Expert box 5.2), but also to support the economy, with many schemes funded through development grants and loans (World Bank, 2014).

As stated previously, there are few transport services whose revenue is sufficient to cover their operating and maintenance costs, and without a subsidy, the fares would be very expensive or unaffordable for low income workers. For example, the New York subway's ticket revenue only covers 45% of its operating costs, with nothing for capital repairs. In contrast, the Hong Kong Mass Transit Railway does make a profit; in 2012 it made a profit of USD 2 billion. Hong Kong is able to do this partly as a result of being a very dense city, but mainly through its model of land value capture, whereby it captures revenue from property that benefits from transport proximity. Singapore is another revenue-positive system (Padukone, N. 2013). The corollary of this is that development creates demand (increased population means increasing numbers of people travelling), and the argument is therefore that the development should contribute towards the costs of the transport systems that serve the development. A good example of how this works is operated by Transport for London (TfL). If new bus services are deemed to be required, TfL will make agreements with the developers to subsidise the new bus services, which are for a fixed term from occupation to cover the costs of buying new buses and associated infrastructure and for running potentially low occupancy services in the early years. The process is transparent and if the new services are financially viable early on, then part of the subsidy is refunded. In other circumstances TfL will negotiate the delivery of new transport infrastructure; this could be a new entrance to the Underground system to ease congestion or a new station. Depending on the level of impact of the development, this may amount to just a contribution to the cost or a complete funding package.

Innovation and improvements to urban public transport systems are happening across the region; the Institute for Transportation and Development Planning (ITDP) awarded the 2015 Sustainable Transport Award jointly to the three Brazilian cities of Belo Horizonte, Rio do Janeiro and São Paulo and the 2017 prize went to Santiago, Chile. These cities were recognised for their progress in transforming their cities by promoting urban regeneration that prioritises pedestrians over cars, bus rapid transit systems, pedestrian and cycle networks, the use of open data to improve access to public transport, the creation of strategic spatial planning policies and the establishment of governance bodies to deliver transport that engage civil society.

Electronic smart card ticketing systems have enormous positive impacts on public transport networks. Firstly, they speed up the boarding process, thus shortening journey times (particularly on buses), they are more environmentally friendly than printing tickets and they reduce litter. They are also seen as a mechanism for encouraging public transport use, as they make journeys more convenient (provided the ticketing system is comprehensive, covering all the transport options and capping the cost of multi-modal journeys) and can be used to create affordable travel for the urban poor. Smart travel cards are now in use in many South American cities, e.g. BIP card in Santiago and the SUBE card in Buenos Aires since 2009 – its use is now planned to be expanded to cover toll roads. The use of these cards is increasingly being introduced in major cities across the region and is being used as a tool to improve and deliver more affordable transportation, as mentioned above.

As South America's population is already largely urban, on average over 80% (see Table 1.5), the challenge in the region is to modify existing cities sustainably, unlike many parts of the developing world that will see high influxes of people to cities that will struggle to cope. Current planning thinking is very much on the side of transit-oriented development (TOD), which suggests that new development should be centred around transport hubs, in an effort to capitalise on existing transport corridors and to prevent urban sprawl. It is therefore important for both planning policy and real estate development to identify the transport hubs within the city and to encourage development around them. Venezuela attempted a decade ago to spread some of its population and manufacturing activity to the inland plains by building the Tinaco-Anaco railway (with Chinese expertise bought in on the back of vast oil revenues) assuming that developers would follow the project. This did not happen and the project has been abandoned, certainly partly due to the collapse of the Venezuelan economy, but also the misguided concept that development would automatically follow the railway.

5.3.2 Energy

The decarbonisation of energy is an essential strategy of climate change mitigation. Models of low carbon living need to be developed, focusing on clean energy sources and making developments more self-reliant in terms of energy. Future developments should not replicate existing carbon-intensive energy solutions. The Intergovernmental Panel on Climate Change (IPCC), the international scientific body on climate change, in its compelling *Special Report on Renewable Energy Source and Climate Change Mitigation*, concluded that close to 80% of the world's energy supply could be met by renewables by 2050, which would cut greenhouse gas emissions by a third (IPCC, 2012). If climate change mitigation targets and the Sustainable Development Goals are to be met, the real estate and development sectors include renewable energy sources or build-in flexibility to transition to these if it is not feasible initially.

Energy infrastructure is expensive, it mostly takes a long time to deliver and is fraught with political uncertainty. Conventional electricity grids are highly inefficient and on average only 2 energy units of electricity are delivered for every 10 units of fossil fuel energy consumed to produce it (Greenpeace, 2006). In the struggle to create more energy-efficient and less fossil-fuel dependent energy systems, new developments should explore different systems for energy supply and delivery, such as decentralised energy.

Good energy strategies should combine energy savings and demand reduction with efficiency, reuse of waste energy and renewable energy sources. There are almost always opportunities to identify symbiotic relationships between different activities; what is waste for one process may be fuel for another, such as the integration of waste plants with local decentralised energy plants. Planning the different aspects of infrastructure should not be done in isolation, but must be thought of holistically, so that interdependencies can be developed to advantage.

5.3.3 Water and sanitation

Water is a valuable and renewable resource and increasing urbanisation is creating ever growing demand. However, in many areas underground aquifers are being drained faster than they are being replenished. Processing water and sewage is energy intensive, yet water is often perceived as having little or no value, especially when it is not charged for at the point of use. As a result, it is often used wastefully and inappropriately. Water strategies in new developments should include water harvesting, water recycling and using grey water, as well as pricing policies and technical restrictions that will help reduce consumption.

Water management regimes divide water into four "colour" categories:

- Blue, which is clean and natural, as it occurs in rivers, lakes, wetlands and aquifers;
- Green, which is water used to support plant growth;
- Grey, which is rainwater runoff from buildings and therefore not entirely pure; and
- Brown, which is sewage and other waste water affected by human activity.

The colour of the water obviously determines its use and value, but it can be improved to a better colour through treatment and filtration processes. New developments should therefore aim to balance demand with quality, to ensure sufficient supply at affordable prices. Buildings need to use and dispose of water efficiently, rainwater needs to be attenuated and captured as far as possible through the inclusion of Sustainable Urban Drainage Systems (SUDS) and landscaping needs to be intelligent and respectful of the local climate.

The problems around water management are wide ranging, from supply of clean drinking water, the treatment of waste water and controlling river pollution to flood prevention. These issues are often interlinked and are managed comprehensively through river basin management or urban water management schemes, that cover large areas (see Expert box 5.2).

5.3.4 Telecommunications

Governments in Latin America have recognised the importance of information and communications technology (ICT) to economic productivity and GDP growth, as well as its role in tackling poverty and social exclusion, by participating in programmes such as the global initiative, *Connect a School, Connect a Community*,[38] aimed at ensuring all schools would be connected to the Internet by 2015 and the *Women's Digital Literacy Campaign*,[39] training Chilean women to use computers, information and communication technology. In Argentina, former President Kirchner launched a programme, *Argentina Conectada*, to tackle access deficiencies by providing broadband internet connections to schools and to develop community knowledge centres, *Núcleos de Acceso al Conocimiento*.[40] The last challenge in most South American countries is to even out the digital divide between rural and urban citizens.

Table 5.9 shows the level of access to telephone and internet in 2016 in South America, though this situation has largely improved since then. The region has

Table 5.9 Internet usage in South America by country

Internet usage for South America, 30 June 2016

South America	Internet usage, 30 June 2016	% population (penetration)
Argentina	34,785,206	79.4%
Bolivia	4,600,000	41.9%
Brazil	139,111,185	67.5%
Chile	14,108,392	79.9%
Colombia	28,475,560	58.6%
Ecuador	13,471,736	83.8%
Falkland Islands	2,800	96.2%
French Guiana	100,000	36.3%
Guyana	305,007	41.4%
Paraguay	3,149,519	45.9%
Peru	18,000,000	58.6%
Suriname	260,000	44.4%
Uruguay	2,400,000	71.6%
Venezuela	18,254,349	61.5%
Total South America	277,023,754	66.7%

Source: Internet World Stats www.internetworldstats.com/stats.htm

seen strong growth in telecommunication access and connectivity, with many countries now having reached 100% mobile phone penetration, though there are still areas of poor access in some rural and urban areas. Brazil leads in the region and is the fifth-largest mobile market in the world and has a number of its own satellites. The Brazilian government has developed a National Broadband Plan, with the goal of providing broadband access to lower-income homes and to areas where private operators have no commercial interest. In 2016, 67.5% of the Brazilian population (some 140 million of the total 208 million population) uses the internet (see Table 5.9).

There has been strong focus on improving ICT across the region; in Colombia, a government initiative has been seeking to close the "digital divide," and provide every Colombian with internet access. Chile was the first country in the world to pass a Network Neutrality Law, to ensure free and equal access to the internet for its citizens.

5.4 Infrastructure funding mechanisms

5.4.1 *Government/municipality funded*

Central government financing is often, though not always, part of the mix of funding packages for major projects and the availability is, of course, dependent upon the economic situation. Central funds are often part of a targeted government policy to tackle a specific problem and may be announced as a high-level growth or poverty alleviation strategy (such as the poverty targeting strategy currently implemented by President Macri in Argentina). These strategies carry heavy political weight and as a result they may fade with administration change. Examples are numerous; London saw major infrastructure projects cancelled when Mayor Boris Johnson was elected as he swiftly cancelled several projects, including the Thames Gateway Bridge and the Cross River Tram.

In May 2016, President Michelle Bachlet of Chile announced plans for an infrastructure fund, to be 99% controlled by the Treasury, that will have assets of about USD 9 billion. Chile's public infrastructure plan includes several government projects aimed at developing the extreme northern and southern regions, which are the ones most affected by recent natural disasters. Included are plans to connect the Southern very remote provinces of Aysén and Magallanes with a submarine fibre optic network to provide more reliable connectivity. In the Northern region, characterised by desert climate, the government aims to launch 20 water projects, including desalination and wastewater treatment plants; it is assumed these projects will have a level of government funding, though they may emerge as part funded through PPP or concession projects (which will have an element of government funding).[41]

President Macri of Argentina is using increasing amounts of public funds for infrastructure projects, in an effort to revive the economy. However, this spending of public money is widening the budget deficit, fuelling inflation and keeping interest rates high.

Brazil introduced central government funding in 2007 through the Growth Acceleration Programme (PAC) to address infrastructure deficiencies. Unlike many other countries, this programme also included funding for operation and maintenance. As is highlighted in Expert box 5.3, since 2010 the vast majority of the PAC funds have gone to the energy sector. The majority of infrastructure projects are either funded by the National Development Bank (BNDES) or through foreign multinational corporations.

Since 2002, Colombia has had legislation allowing national government to fund up to 70% of capital costs for transport projects, with local authorities contributing the balance, designed to encourage efficient public transport systems (see Expert box 5.2), which has helped cities move towards integrated and efficient transport systems (CONPES, 2002). It has seen some problems such as local authorities' capacity to raise match-funding and consequent funding reviews. Other problems have been associated with relocation of utilities networks and difficulties acquiring land. However, the programme has delivered new and renovated road infrastructure, pedestrian public spaces, bike paths, bus shelters infrastructure and traffic control systems and, in larger cities, it has prompted further development of BRT and cable car systems and recently it has allocated funding for the construction of Bogota's first subway line (Johnson, 2017).

This funding mechanism has resulted in some of Colombia's cities now seeing significant improvements to public transport networks, Medellín being a prime example. The city has developed an exemplary approach to using transport infrastructure to promote social inclusion and equality by delivering access to its public transport system, even for the inhabitants of the informal settlements on the hills above the city, by integrating cable car systems with the public transport network. The system provides access to opportunity and work for thousands of residents that previously had very long commutes to access work in the city centre.

The cable car initiative is an exemplar of best practice in regeneration, urban renewal and social progress and has been the recipient of many awards. It was developed with community participation and has resulted in creating accessible neighbourhoods that are no longer torn by rival gang cultures. The cable car system also integrated a network of public libraries giving the residents access to books and information. The scheme was funded 100% through local government financing, so it did not apply for the national government funding.

Reduction in public expenditure in Peru in recent years has slowed down the rate of growth of the construction sector. Local and regional public investment fell 40.2% and 31.3%, respectively, in the first half of 2015. Collectively, local and regional public investment account for 63% of total public investment in the country. The impact of this decrease was only partly offset by a 16.4% increase in investment by the central government over the same period (Miranda de Sousa Hernández-Mora, 2016)

5.4.2 Private investment (international and local) and government bonds

The World Bank (2016) has highlighted that the participation of the private sector is a fundamental and increasingly important aspect of delivering

Figure 5.1 Metrocable, Medellín
© Camilla Ween

infrastructure as this is a way to ease the burden of finance for governments, as well as avoiding the peaks and troughs of GDP performance. Partnering can also reduce operational subsidies for governments by providing funds when they are not available to the government and have the potential to allow delivery to be more responsive to public need. However, the World Bank does point out that these arrangements require compliance from both sides and should have an impartial regulatory body to monitor progress and outcomes.

Because infrastructure projects have long timescales and pension funds are able to invest for long periods, investment in infrastructure by pension funds can be a good match. However, pension funds have a mandate for high returns in order to guarantee pensions to its members. For this reason many tend to shy away from high risk investment such as infrastructure in emerging countries. A study by BBVA Research in 2012 highlighted this synergy and looked at investment in Latin America by pension funds. Table 5.10 shows that pension funds are investing in infrastructure, particularly in Brazil (USD 62.6 billion), Colombia (USD 9.5 billion), Chile (USD 14.5 billion), and Peru (USD 3.4 billion). However, the majority of the investment has been channeled into energy projects. It remains to be seen if pension funds will play a significant role in financing other infrastructure projects in South America.

The rise of private financing in infrastructure has been reshaping infrastructure financing in Peru and private initiative projects have become increasingly common, particularly for energy and water projects.[42] New regulatory instruments for private-public cooperation are being created. Also, a "Public Works for Taxes" scheme is increasingly being used by private companies to contribute

Table 5.10 Level of investment by pension funds in Latin America

Infrastructure investment by pension funds in South America		
	US$ billions	*As % of GDP*
Brazil	62.6	3.0%
Colombia	9.5	3.5%
Chile	14.5	6.5%
Peru	3.4	2.3%

Source: BBVA Research, 2012

to their local communities and for the government to attract more private financing for infrastructure. Businesses can offset certain tax expenses by building infrastructure such as roads, schools and health care facilities in the areas close to their operations; the scheme has focused mainly on large-scale extraction firms working in the provinces. The scheme yielded over USD 510 million for infrastructure between 2009 and 2015. This has forced efficiency on the construction industry and led to more competitive tendering (Miranda de Sousa Hernández-Mora, 2016).

In Brazil, private investment in infrastructure is mainly through foreign multinational corporations. Local private sector investment is currently very low due to the poor capital market situation.

Argentina expects that large scale projects will rely on international multinationals. For example, it has attracted USD 5.6 billion of private sector investment from German Seimens AG, partly, backed by Germany's export credit agency, to finance high-efficiency gas power plants and wind-generated renewable energy, energy efficiency in buildings and rail automation.[43] In Argentina cities, municipalities and provinces are raising finance for infrastructure projects through bond issues. President Macri offered a tax amnesty in 2016, in the hope of boosting investment in infrastructure projects.

Chile has strong institutional frameworks, high levels of legal transparency and low levels of corruption, which offer strong financial and legal safeguards for private sector investors.

Chile also employs the concession model for infrastructure delivery, which is a means of levering in private finance. An example is the Punilla Reservoir on the Ñuble river basin, which also includes a hydroelectric plant, and which will require the concessionaire to design, construct and operate the facility; the investment value is USD 387 million.[44]

5.4.3 Public Private Partnership (PPP)

The report *Regenerating Urban Land: A Practitioner's Guide to Leveraging Private Investment* (World Bank, 2016) claims that private sector participation is the single most crucial component in rejuvenating decaying urban areas around the world. It is increasingly difficult for local or national governments to completely fund infrastructure projects. Building on the experience of cities around the world, the report looks at a variety of schemes; what they all have in common is significant private sector participation in the regeneration and rehabilitation of deteriorating urban areas. The report singles out successful policy and finance tools and points out issues and challenges each city faced during the process, but stresses that the successful use of land-planning and finance tools depends on sound and well-enforced policy and property tax systems.[45]

However, Professor David Hall of the University of Greenwich is less sanguine about the value of PPPs (Hall, 2014). Using empirical evidence, Hall suggests that PPPs do not deliver the benefits that are promised; that they come with systemic problems of corruption, secrecy and unreliable forecasts, and that they have a damaging effect on public services. Even the World Bank (2016) in its outline of the pros and cons of PPPs, draws attention to issues around false expectations that the public sector does not have to pay for services, and that over time, the reality is that a PPP will mean higher public spending than with conventional projects.

In an attempt to shift the burden of funding away from central government, Brazil introduced what is effectively a Public Private Partnership (PPP) growth programme, *Projeto Crescer* (Project Growth) in 2016 that was aimed at attracting private sector driven investment.[46] The government launched a round of privatisations that came with associated access to credit from the national development bank BNDES.

Peru's new administration is looking to increase infrastructure spending and develop the role of PPPs to promote economic growth. President Pedro Pablo Kuczynski has pledged to boost infrastructure development through some 80,000 large and small projects spanning various sectors from civic buildings to the construction of dams, roads and railways. To achieve this, the new administration is reviving and decentralising *ProInversión*, the Private Investment Promotion Agency of Peru, to facilitate increased regional independence and ease project implementation.[47] Each region will be expected to have its own investment plan to encourage more participation from the local population and promote transparency (Miranda de Sousa Hernández-Mora, 2016).

In 2014, Peru spent 4.46% of GDP on infrastructure compared to a regional average of 3.4% (see Table 5.2). PPPs are seen to offer a way for the private sector to participate in accelerating the delivery of Peru's much-needed infrastructure development as well as allowing foreign capital to finance large-scale projects. President Kuczynski has pledged to fast-track 15–20 previously delayed projects by using PPP financing mechanisms. These include Lima's Metro 2 line, a natural gas pipeline running across the Cusco region, the expansion of Jorge Chavez International airport in Lima, the development of the country's second international airport, Chinchero International, and the Lima-Ica highway. Together, these projects represent USD 18 billion in investment (www.ProInversión.gob.pe). The 35 km Metro 2 line, expected to cost over USD 5 billion, will be one of the biggest PPP contracts and will include World Bank funding. The project is projected to complete in 2021. Investment is also planned for airports and developing a new natural gas pipe line to boost energy supply under PPP arrangements. An often cited problem with getting large projects off the ground is the local lack of skills (see Expert box 5.2). Closer collaboration with universities and potential secondment of staff or master students would be one option, but *ProInversión* has been providing regional seminars and training for state employees to address the skills gap.

In Argentina, under the previous regime of the Kirchners, foreign direct investment lagged behind that of Brazil, Mexico, Chile and Colombia and at times Peru, even though Argentina is the region's third-largest economy.[48] To lure investors, the current government has drafted a PPP bill, a system hardly used in Argentina, to renew the country's infrastructure. The goal is to attract more than USD 5 billion a year in investments through these partnerships (Newbery, 2016).

Chile introduced in 2016 the Foreign Investment Act to regularise foreign investment, such as PPPs and concessions.[49] The PPP model is being employed for some of its infrastructure financing, such as the Costanera Norte highway project, which consists of a six-lane highway, two tunnels and six bridges, as well as over 30 km of urban highway. As stated, Chile's economy is stable and with a reputable PPP process now in place, it is likely to be able to attract investment and participation in PPP projects. Concessions are another mechanism being used for raising finance, such as the aforementioned Punilla Reservoir (see Section 5.2).

5.4.4 Multilateral development banks (MDB) and investment banks

The main multilateral development banks operating in South America are the World Bank, the Inter-American Development Bank, Andean Development Corporation or Development Bank of Latin America (CAF), the Chinese Development Bank and the Green Climate Fund. Table 5.11 shows that in 2014 all countries in South America were relying to some extent upon MDB funds.

Table 5.11 Investment by multilateral development banks in South America

World Bank Group						Inter-American Development Bank	
IBRD 2015		*IDA*		*IFC*		*IDB Group*	
Colombia	1,400	Bolivia	100	Brazil	1,505	Brazil	2,241
Argentina	1,337			Chile	389	Colombia	932
						Argentina	660
						Peru	445
						Uruguay	343
						Paraguay	301

Notes: US$ millions

International Bank for Reconstruction and Development (IBRD)

Inter-American Development Bank (IDA)

International Finance Corporation (IFA)

Inter-American Development Bank (IDB)

Source: MDB Annual Reports, 2014

Between 1997 and 2001, Argentina and Brazil, together with Mexico, accounted for one-third of the World Bank's total global commitments (Avalle, 2005).

The World Bank has been investing in mass transit corridors in South America for some years, such as the Lima and Colombian BRT systems and the Brazilian *Companhia Brasileira de Trens Urbanos* – CBTU (Brazil's urban rail company). It is now also investing in metro systems, such as the São Paulo Metro line 4, which was the first metro to be part-funded by the World Bank. It has also been investing in road access to the favelas in Brazil. In Santiago, the World Bank has been closely involved in the upgrade of the city's urban transport system, *Transantiago*, which includes bus infrastructure and improved interchange. The Bank supported the implementation of the Santiago Urban Transport Plan through a technical assistance loan. In 2017, the World Bank confirmed a loan for Argentina of USD 45 million for a Metrobus BRT system and the Sáenz Transfer Centre in Buena Aires.

The Development Bank of Latin America (CAF) was established in 1970, comprising 17 Latin America and Caribbean countries and Spain and Portugal, together with 14 private banks. It promotes sustainable development through credit arrangements, grants and support for technical and project structuring. In Argentina CAF is funding a wide range of environmental and transport Projects, such as the flood control plan for the Lujan River Basin and the Belgrano Sur Passenger Rail project (see Expert box 5.2). Argentina is also relying on multilateral lenders like the World Bank funds for long-term projects such as transmission lines (Newbery, 2016).

The year 2015 saw a number of European and international banks shrinking operations or pulling out and closing branches in South America, mainly due to the region's poor performance and pressure in Europe to discard weak businesses. The key players included French Société Général, German Deutsche Bank AG, HSBC and Barclays Plc, though it is anticipated that others such as Spain, Switzerland, and the US would be willing to wait out the economic downturn (see Graphic 5.2). In the case of Deutsche Bank, its pulling out had more to do with its own financial situation and having to meet fines over its involvement in the Libor scandal, rather than the regional circumstances. Clearly the dwindling of lending sources is putting further pressure on governments and institutions to raise finance for projects.

This was a few months after Deutsche Bank announced a finance package for a USD 144 million renewable energy onshore wind farm in Uruguay.[50] Deutsche Bank, acting as the structure bank, also brought in Commercial Bank of China.[51]

For 2015 and 2019, China's President Xi Jinping has set a high target of USD 250 billion in direct investment in the Latin American and Caribbean region (LAC). The pledge was made in January 2015 at the first ministerial meeting of the Forum of China and the Community of Latin American and Caribbean States, which was held in Beijing. As seen in Table 1.1, China is already investing heavily in the region and it is expected that in the next few years, the Asian country is likely to emerge as the world's largest supplier of capital and a key investor in South America.

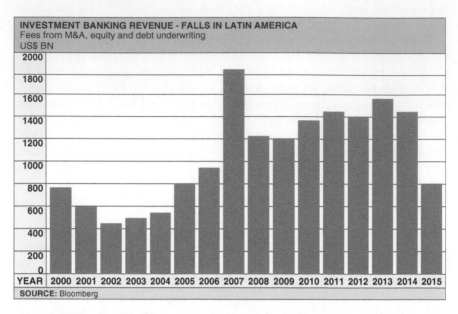

Graphic 5.2 Investment banking falls in Latin America

5.5 Evaluation of infrastructure development approaches in the Big Five

One of the key challenges for delivering sustainable development in South America will be the funding of strategic infrastructure. The level of investment in infrastructure remains below the 5% of GDP level deemed necessary (see Expert box 5.2) and needs to rise; otherwise, unsustainable development will be commonplace. However, if developers work closely with governments and funding institutions to ease up access to finance and develop potential long-term payback mechanisms (for example by mortgaging future land value capture and taxes towards repayment of finance) then greater certainty over the delivery of infrastructure should be possible.

The poor track record of investment in infrastructure in Brazil has been partly blamed on uncertainty surrounding delay and due to the regulatory process (see Expert box 5.1). However, the current *Projeto Crescer* has introduced greater emphasis on reform of the regulatory environment. Other problems cited as having contributed to the poor delivery of infrastructure are the high costs of licenses and the awarding of energy contracts to those offering the lowest eventual user tariffs (which may not be sustainable in the long term). Though Brazil has raised the percentage of GDP invested by government funding in infrastructure (up from 2% prior to 2007 to 3.2% in 2014), the vast majority of investment since 2010 (92.4%) has been focused on the energy sector. Public transport (as well as water and sanitation) infrastructure investment in Brazil is woefully lacking, and as a result the emphasis has been on highway construction, with the resulting consequences of very high car use, traffic congestion and, as it will be shown in the next chapter, poor air quality. If more integrated land use and transportation policies emerge over the coming decades, then one would hope to see more transit-oriented development and the use of land value uplift tools could help to cross-fund public transport as well as other infrastructure.

Colombia differs from many developing countries in that the national government works with local governments to invest in public transport infrastructure, which has helped bring about the provision of affordable transit; an essential aspect of sustainable mobility within cities. Even when funding opportunities exist, many cities do not have the capacity or skills to deliver complex schemes, due to the best universities offering the appropriate training are not located in medium sized cities. A priority should be to establish agencies that can assist in the delivery of programmes; these could be developed through collaboration between government and universities. Money will ultimately be well invested if the people leading delivery programmes, contracting, building and designing further phases are located on the ground in each city.

Investment in infrastructure in Chile is on the rise and with President Bachelet's Chile 30.30 investment programme one can expect infrastructure improvements across the sectors. However, as the Expert box 5.3 highlights in relation to transport infrastructure in Santiago, the targeting of investment could be better aligned to user demand, with less emphasis on highways and car travel.

Argentina, which saw strong private investment in the 1990s, has seen this fall noticeably since the start of the new millennium, but percentage of GDP investment has overall been rising over the last few years (see Expert box 5.2). Significant transport and water management infrastructure projects are currently in the pipeline.

Notes

1 World Economic Forum's Global Competitiveness Report. (2016). Available at www3. weforum.org/docs/gcr/2015-2016/Global_Competitiveness_Report_2015-2016.pdf, accessed on 10th July 2017.
2 (2017) *Reuters World News*.
3 (2017) *Government of Peru: ProInversión.gob.pe*.
4 As reported by Financial Times, *Brazil's Waterways Could Decongest Gridlocked Roads*. Available at www.ft.com/content/7b714ce6-f2b3-11e2-a203-00144feabdc0?mhq5j=e1, accessed on 10th July 2017.
5 Meyer, A. (2010) *Brazil Infrastructure 2010*.
6 World Bank (2012).
7 *Ibid.*
8 *Ibid.*, p. 78.
9 *Ibid.*, p. 79.
10 Mourougane and Pisu (2011, p. 25).
11 *Ibid.*
12 *Ibid.*, p. 10.
13 Investorideas.com, 20th February 2014. Available at www.investorideas.com/news/ 2014/international/02204.asp
14 As reported by Financial Times, *Argentina Turns to Renewable Energy*. Available at www.ft.com/content/c6e58576-2da1-11e6-bf8d-26294ad519fc?mhq5j=e1, accessed on 10th July 2017.

15 See http://infralatam.info/ – joint initiative by CAF, IADB and Cepal to measure public and private infrastructure investments in Latin American countries.
16 Bhattacharya, A., Romani, M., and Stern, N. (2012) *Infrastructure for Development: Meeting the Challenge*. London: Centre for Climate Change Economics and Policy. Available at www.cccep.ac.uk/wp-content/uploads/2015/10/PP-infrastructure-for-development-meeting-the-challenge.pdf

   CAF, banco de desarrollo de América Latina. (2011) *La infraestructura en el desarrollo integral de América Latina*. Bogotá: IDeAL.

   Calderón, C. and Servén, L. (2003) "The Output Cost of Latin America's Infrastructure Gap." In *The Limits of Stabilization: Infrastructure, Public Deficits and Growth in Latin America*, ed. W. Easterly and L. Servén, 95–118. Stanford, CA: Stanford University Press.

   ECLAC (Economic Commission for Latin America and the Caribbean). (2010) "The Economic Infrastructure Gap in Latin America and the Caribbean." In *Facilitation of Transport and Trade in Latin America and the Caribbean*, 332, 4/2014, p. 5.

   Serebrisky, T., Suárez-Alemán, A., Margot, D., and Ramirez, M. (2015) *Financing Infrastructure in Latin America and the Caribbean: How, How Much and By Whom?* Washington, DC: Inter-American Development Bank.
17 Calderón, C. and Servén, L. (2010) Infrastructure in Latin America. *Policy Research Working Paper 5317*. Washington, DC: World Bank.
18 Standard and Poor's. (2015) *Global Infrastructure Investment: Timing Is Everything (and Now Is the Time)*. Standard & Poor's Ratings Services, McGraw-Hill Financial. Available at www.tfreview.com/sites/default/files/SP_Economic%20Research_Global%20Infrastructure%20Investment%20(2).pdf
19 CAF is a development bank established in 1970, comprising 19 countries – 17 from Latin America and the Caribbean plus Spain and Portugal – and 14 private banks, also from the region. It promotes a sustainable development model through credit transactions, grants and support for technical and financial project structuring for public and private sectors in Latin America.
20 Information retrieved from the Project Economic Appraisal Document. CAF, 2016.
21 *Ibid.*
22 Szenkman, P. (2015) Menos autos y más y mejor transporte público para la Región Metropolitana de Buenos Aires. *Centro de Implementación de Políticas Públicas para la Equidad y el Crecimiento – CIPPEC*. Buenos Aires. Available at www.cippec.org/documents/10179/51825/149+DPP+ADE,%20Menos+autos+y+m%C3%A1s+y+mejor+transporte+p%C3%BAblico+para+la+Regi%C3%B3n+Metropolitana+de+Buenos+Aires,%20Szenkman+2015.pdf/afb097fb-5f95-4301-bc2e-9f950f118a30
23 This is Chile, *Business, Daily Life, Tourism News*. Available at www.thisischile.cl/transportation-and-infrastructure/?lang=en, accessed on 10th July 2017.
24 Chile Presents US$28bn Infrastructure Master Plan, *BN AMERICAS*. Available at www.bnamericas.com/en/news/bachelet-rolls-out-connectivity-infrastructure-projects-for-southern-chile/?position=648184, accessed on 10th July 2017.
25 (8th December 2016).
26 Turning Chile's Infrastructure Challenges Into Opportunities, *Infrastructure Intelligence*. Available at www.infrastructure-intelligence.com/article/sep-2016/turning-chile's-infrastructure-challenges-opportunities, accessed on 10th July 2017.
27 Infrastructure Investment in Chile is on the Rise, Supported by the Government's Growth Agenda, *Timetrics Infrastructure Intelligence Centre*. Available at www.infrastructure-ic.com/pressrelease/infrastructure-investment-in-chile-is-on-the-rise-supported-by-the-governments-growth-agenda-5684038, accessed on 10th July 2017.
28 A term from sociology to describe the interplay between a set of values, symbols, cultural norms, laws and institutions usually collectively held, by specific groups or a society as a whole.
29 International Trade Centre.
30 CONPES 3167, Departamento Nacional de Planeación de Colombia
31 CONPES 3833, Departamento Nacional de Planeación de Colombia.

32 *Ibid.*
33 This figure does not include the Bogota program.
34 (23rd May 2015). *BBC News.*
35 (2006) *International Energy Agency.*
36 The Guardian, *Uruguay Makes Dramatic Shift to Nearly 95% Electricity From Clean Energy.* Available at www.theguardian.com/environment/2015/dec/03/uruguay-makes-dramatic-shift-to-nearly-95-clean-energy, accessed on 10th July 2017.
37 Rebelo, M. (2013) *World Bank.*
38 See programme at www.itu.int/en/ITU-D/Digital-Inclusion/Youth-and-Children/Pages/CSCC.aspx
39 See programme at www.itu.int/en/ITU-D/Digital-Inclusion/Women-and-Girls/Pages/Digital-Literacy.aspx
40 See programme at www.itu.int/net/itunews/issues/2011/07/24.aspx
41 *Ibid.* endnote 10.
42 Time to Build, Construction and Real Estate, Peru, *The Business Year.* Available at www.thebusinessyear.com/peru-2017/time-to-build/review, accessed on 10th July 2017.
43 Siemens to Run $56 Billion in Argentina Infrastructure Projects, *Bloomberg L.P.* Available at www.bloomberg.com/news/articles/2016-09-14/siemens-agrees-to-5-6-billion-argentina-infrastructure-project, accessed on 10th July 2017.
44 *Ibid.* endnote 10.
45 Full Report and Toolkit. Available at http://urban-regeneration.worldbank.org
46 www.projetocrescer.gov.br
47 www.proinversion.gob.pe
48 (2016) *United Nations Economic Commission for Latin America and the Caribbean (ECLAC).*
49 *Ibid.* endnote 10.
50 (2015) *Bloomberg.*
51 (2015) Deutsche Bank Finances Its First Renewable Energy Project in Latin America, *Deutsche Bank.* Available at www.db.com

Bibliography

Avalle, O. (2005) The Multilateral Development Banks in Latin America and the Caribbean Region. *Vermont Journal of Law*, 6, pp. 196–199.

Bird, K., McKay, A., and Shinyekwa, I. (2010) *Isolation and Poverty. The Relationship Between Spatially Differentiated Access to Goods and Services and Poverty.* Overseas Development Institute. Available at www.odi.org/sites/odi.org.uk/files/odi-assets/publications-opinion-files/5516.pdf, accessed on 10th July 2017.

Calderón, C., Moral-Benito, E., Servén, L. (2011) Is Infrastructure Capital Productive? A Dynamic Heterogeneous Approach. *Policy Research Working Paper 5682.* World Bank.

CONPES – Consejo Nacional de Política Económica y Social. (2002) *Política Para Mejorar el Servicio de Transporte Públic Urbano de Pasajeros, Document 3167, Departamento Nacional de Planeación de Colombia.* Available at https://colaboracion.dnp.gov.co/CDT/Conpes/Económicos/3167.pdf, accessed on 10th July 2017.

Deutsche Bank. (2015) *DB Finances its First Renewable Energy Project in Latin America.* Available at https://www.db.com/cr/en/concrete-deutsche-bank-finances-renewable-energy-project-in-latin-america.htm?kid=responsibility.inter-ghpen.headline

Greenpeace. (2006) *Decentralising Energy: Cleaner, Cheaper, More Secure Energy for 21st Century Britain.* Available at http://cut-trident.greenpeace.org.uk/media/reports/decentralising-uk-energy, accessed on 10th July 2017.

Hall, D. (2014) *Why Public-Private Partnerships Don't Work: The Many Advantages of Pubic Alternative.* Available at www.world-psi.org/sites/default/files/rapport_eng_56pages_a4_lr.pdf, accessed on 10th July 2017.

Harvey, J. and Jowsey, E. (2004) *Urban Land Economics, Sixth Edition*. London: Palgrave.

International Energy Agency. (2006) *World Energy Outlook 2006*. Available at www.iea. org/publications/freepublications/publication/weo-2006.html

IPCC – Intergovernmental Panel on Climate Change. (2012) *Renewable Energy Sources and Climate Change Mitigation*. Available at www.ipcc.ch/pdf/special-reports/srex/SREX_ Full_Report.pdf, accessed on 10th July 2017.

ITF – International Transport Forum. (2010) *Reducing Transport Greenhouse Gas Emissions. Trends and Data 2010*. Available at www.itf-oecd.org/sites/default/files/docs/10ghgtrends. pdf, accessed on 10th July 2017.

Lee, B. (2015) China, Brazil, Peru Eye Transcontinental Railway Megaproject. *International Business Times*. Available at www.ibtimes.com/china-brazil-peru-eye-transcontinental-railway-megaproject-1930003

Meyer, A. (2010) *Brazil Infrastructure*. Available at www.brazil.org.za/brazil-infrastructure. html

Miranda de Sousa Hernández-Mora, R. (2016) *Peru Plans to Boost Infrastructure*. Available at www.linkedin.com/pulse/peru-plans-boost-infrastructure-development-ricardo-miranda-de-sousa-, accessed on 10th July 2017.

Newbery, C. (September 8, 2016) Will Infrastructure Investments Revive Argentina's Economy? *Latin Finance*. Available at www.latinfinance.com/Article/3584224/Will-infrastructure-investments-revive-Argentinas-economy.html#/.WWNpoMbMzBI, accessed on 10th July 2017.

Padukone, N. (2013) The Unique Genius of Hong Kong's Public Transportation System. *The Atlantic*. Available at www.theatlantic.com/china/archive/2013/09/the-unique-genius-of-hong-kongs-public-transportation-system/279528

Ramos, D. (2017). German Companies Interested in Train Crossing South America: officials. *Reuters*. Available at www.reuters.com/article/us-bolivia-railways/german-companies-interested-in-train-crossing-south-america-officials-idUSKBN16U00M

Rebelo, J.M. (2013) *Urban Transport in Latin America*. World Bank presentation. Available at www.youtube.com/watch?v=GqRn5JweZDU

Stern, N. (2006) *The Economics of Climate Change*. HM Treasury and Cabinet Office, Full Review. Available at http://webarchive.nationalarchives.gov.uk/+/www.hm-treasury. gov.uk/sternreview_index.htm.

Taj, M. (2017) Abnormal El Nino in Peru Unleashes Deadly Downpours; More Flooding Seen. *Reuters*. Available at www.reuters.com/article/us-peru-floods/abnormal-el-nino-in-peru-unleashes-deadly-downpours-more-flooding-seen-idUSKBN16O2V5

UN Habitat. (2011) *Hot Cities: The Battle-ground for Climate Change*. Available at http:// mirror.unhabitat.org/downloads/docs/E_Hot_Cities.pdf, accessed on 10th July 2017.

WEF – World Economic Forum. (2016) *Renewable Infrastructure Investment Handbook: A Guide for Institutional Investors*. Available at www3.weforum.org/docs/WEF_Renewable_ Infrastructure_Investment_Handbook.pdf, accessed on 10th July 2017.

World Bank. (2014) *Targeted Subsidies in Public Transport: Combining Affordability With Financial Sustainability*. Available at http://siteresources.worldbank.org/INTURBAN-TRANSPORT/Resources/340136-1152550025185/Targeted-Subsidies-Public-Trans port-Note-04-23-2014.pdf, accessed on 10th July 2017.

World Bank. (2016) *Regenerating Urban Land: A Practitioner's Guide to Leveraging Private Investment*. World Bank and the Public Private Infrastructure Advisory Facility (PPIAF). Available at www.worldbank.org/en/topic/urbandevelopment/publication/ regenerating-urban-land-a-practitioners-guide-to-leveraging-private-investment, accessed on 10th July 2017.

6 Barriers to cross-border investment in South America

6.0 Introduction

Since the late 1980s, investors have greatly diversified their assets thanks to the advent of new technologies that facilitated the transfer of funds from country to country (Garrett, 2000; Talalay, 2000; Sassen, 2006). This has in turn increased their appetite for profit, fuelling a global boom in international institutional financial operations which is helping to push down barriers to foreign direct investment (FDI)[1] (Baum and Murray, 2010). In the case of real estate, financial globalisation has also helped to create new investment vehicles such as Real Estate Investment Trusts (REITs), which have helped to solve liquidity issues that are characteristic of the real estate asset class (see Chapter 4 and in particular Expert box 4.1 for details of REIT-like structures available in South America).

The benefits of FDI to developing countries has been the topic of considerable academic debate since the late 1960s (Dunning, 1994). Some of the issues relate to the externalities that host countries are likely to receive with FDI. For instance, foreign firms looking to expand into other markets might be able to bring in new technologies to the host country, whereas others argue that this is not always the case, as certain conditions must exist in the host country for the externalities to occur (Loungani and Razin, 2001). As the academic work surrounding this debate is considerable, the following paragraph provides the two sides of the argument that relates to the countries under study in this volume.

From the 1960s to the late 1990s, Latin America along with East Asia had been the dominant recipients of most of the world's FDI flows to developing countries, reaching 92% during the last decade of the previous century (Zhang, 2001). For this research, Zhang considers 11 countries around the world including Argentina, Brazil and Colombia in South America, as well as Malaysia and Singapore, two of the Asian Tigers that have been the recipients of a considerable amount of real estate FDI flows in recent years (GRETI, 2017). Yet as Zhang suggests, the East Asian nations during the 1960s were, relative to Latin America, much poorer nations with fewer economic prospects. The author concludes that the benefit of FDI is country specific, and that having a liberalised trade regime, improved education and being able to maintain macroeconomic stability helps countries to reap the benefits of FDI. This coincides with other studies which show that many countries in Latin America can benefit from FDI in, for example, education and transport projects (Bengoa and Sanchez-Robles, 2003). Although, along with Zhang, these authors point out that benefits from FDI can only be achieved if countries commit to improving political and economic stability and can guarantee economic freedom to investors. On the side of the critics, the most compelling argument against the benefits of FDI has been presented in a case study of Venezuela (Aitken and Harrison, 1999) where the authors use data for the period of 1976 to 1989, a time when the country received a considerable inflow of FDI. After revising more than 4,000 businesses in different sectors, the authors found no evidence that FDI could benefit the local economy.

Government policies on FDI vary in South America. For instance, policies followed by Argentina during the Néstor Kirchner administration (2003–2007), consisting of selective protectionism and targeted state intervention, were certainly a response to the increased globalisation and liberalisation of markets seen in the region during the 1990s (Wylde, 2011; see also Expert box 5.4 for a point of view of the impact of liberalisation of markets in Chile). Stances like that of the Kirchner administration have prompted criticisms from international organisations that maintain that the region has more room for improvement and that the financial markets in Latin America as a whole can be greatly deepened if the region were to move towards reducing certain restrictions that are hindering integration with global markets (IMF, 2016). The same source highlights the case of Brazilian commercial bank Itaú. This bank has a wide regional approach for its expansionary policy, but the expansion of other countries into Brazil is hindered, as other similar organisations from the region find it difficult to enter the Brazilian banking sector, mainly because doing so requires presidential approval. Equally, the majority of pension funds' assets in Latin America are very restricted as to how much they can invest abroad. This makes financing large infrastructure projects difficult, as domestic capital markets are insufficient to provide efficient investment opportunities (see Expert box 5.2 for a criticism on the lack of available funds for infrastructure investment in the region). For some years now, Mercosur (see Chapter 1) has been a vehicle for integration amongst Argentina, Brazil and Uruguay, but the IMF suggests that further integration is needed. However, it must be pointed out that economists tend to favour FDI and push emerging nations towards liberalisation of the markets and global integration (Loungani and Razin, 2001), so IMF's recommendations for the region should be considered in the context of the debate described above on the benefits that FDI can bring to emerging markets. In other words, with caution.

The same cautionary approach is applied by investors. From their point of view, the potential benefits of higher returns in investing in another country come at a cost of increased complexity of execution of the operation; therefore, home bias remains an observable phenomenon (Imazeki and Gallimore, 2010). The authors refer to "barriers" and to "deadweight costs" that inhibit an optimal diversification strategy for investors. This excess cost is a loss of economic efficiency that can be caused by country specific factors such as inflation, taxes or monopoly pricing and disinformation (see also Expert box 6.1, and for issues related to lack of appropriate regulation see Expert box 4.1). As a result of this deadweight cost, some countries attract less capital than others, which means that large and more advanced economies will always dominate real estate FDI. For instance, in the case of real estate and according to JLL, 70% of global assets are located in Anglo-Saxon countries (GRETI, 2016). Given the relevance of real estate investment as a driver of economic development in emerging markets (Lapoza, 2007), a levelling-out of economic prosperity may be inhibited by the deadweight costs or barriers to FDI. South American governments should, therefore, be concerned in understanding the barriers to cross-border real estate investment, both real and perceived, for the benefit of investors seeking diversification and return, and for the benefit of governments seeking to promote domestic economic development.

To comprehend the extent of the region's shortcomings in real estate FDI, Expert box 6.1 presents historical inflows of foreign capital received by the entire Latin American region for the past ten years. Using data collected by Real Capital Analytics, which tracks transactions above USD 10 million, the section provides clear evidence of the potential that the region can have for investors. To complement this perspective, the remainder of this chapter will study more closely the deadweight cost to investors in South America, by following the list of real estate barriers uncovered by Baum and Murray (2010).[2] The barriers are analysed by using different global indices, such as the Global Real Estate Transparency Index (GRETI) produced by JLL[3] and the Economic Freedom Index (EFI) produced by the Heritage Foundation.[4] For each index, a component that can best illustrate by proxy a particular barrier will be selected and contrasted with high-performing Anglo-Saxon nations or the Asian Tigers mentioned above.

Expert box 6.1   Issues and opportunities in cross-border investments in Latin America

Jim Costello

Cross-border investment activity is a growing story worldwide but most of the capital moving from one country or region to another is focused on North America, Europe and Asia. Latin America is often not on the target list of investors. There are a number of reasons why these investors shy away from Latin America, but the experience of those investors who do put capital to work in Latin America can be instructive.

In this section we will show that the investment activity in Latin America by investors outside the region delivers two important characteristics. First, the region can provide scale for investors from wealthy countries with few investible assets. Second, investments in the region can provide much higher average yields than other parts of the global property market. Despite these characteristics, this region is far less liquid than others with less transaction activity relative to the size of the economy.

Why is Latin America under-represented?

Cross-border investment activity has been growing as a share of all global commercial real estate investment since 2012. There are a number of forces driving this growth including growing wealth in the developing economies, petro wealth being recycled into productive investments and capital more easily chasing higher yields given better information availability about the market place.

This cross-border activity hit a near-term peak in Q3'15 representing 25% of all deal activity worldwide. In US dollar terms, this activity totaled $357 billion of transactions priced at $10 million and greater. Of all of this cross-border activity, Latin America, as a destination for capital, represented only 1% of cross-border activity worldwide. This is despite an economy that is 7% of the global economy.[5]

All other things equal, one might expect that the larger the economy, the more need there would be for commercial real estate to service that economy. That is the final use of commercial real estate after all: a box in which we house the economy. The larger the economy, to a point, one might expect a greater need for commercial real estate. This said, investment activity in Latin America does not transact at its "fair share" of the global economy.

This disconnect is not a function of an underdeveloped regional economy that does not require commercial real estate. There are thriving, modern metropolitan areas across the region as well as backwater locales focused on primary economic activity such as farming and forestry.

Graphic 6.1 While cross-border investment grows, Latin America is not a major destination

Source: Real Capital Analytics.

Across the 15 countries for which commercial property transaction data is available, there is a wide range of economic productivity. The tourism-based economies of Barbados, Trinidad and Tobago and others not shown[6] in the chart in Graphic 6.2 have the highest rates of economic productivity as measured by US$ GDP per capita, yet lower productivity economies such as Brazil and Mexico have vastly higher investment activity.

The Caribbean countries analysed have less than 1% of the population of Brazil though, so to see if the higher productivity plays a role in higher deal activity, the chart places on the y-axis the deal volume per capita so as to neutralise the impacts of market size.

There is, however, no meaningful relationship in place though with countries such as Argentina, Panama and Chile exhibiting similar levels of productivity but Chile leading the others for both absolute and relative deal volume. The biggest disconnect is between Mexico and Brazil, with Mexico posting about $1 billion more in deal volume from 2012–2016 than Brazil with much higher deal volume per capita.

There are cultural, legal and economic variations across Latin America driving variation in investment activity.

In Brazil, commercial property investment has been on a down cycle in the period shown in Graphic 6.2. as measured in USD. Deal volume hit a peak annual level of $11.7 billion in 2011, but in the 12 months to June 2016, activity totaled only $2.2 billion. The global economic recovery, commodities boom and growing exports that fueled the real estate surge into 2012 just are not in place to help Brazil today.

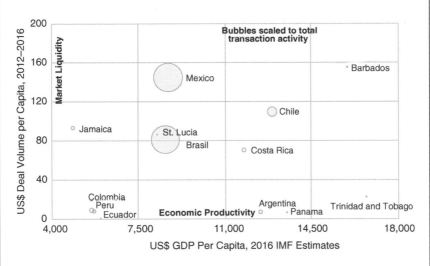

Graphic 6.2 Market breadth and economic productivity not correlated as one might expect

Sources: Real Capital Analytics, IMF World Economic Outlook October 2016

In Mexico, by contrast, USD-denominated commercial property investment grew into 2014 as it pulled back elsewhere. The story in Mexico was one of domestic forces and economic liberalization. RCA's coverage of Mexico in 2008 started at a time of turmoil in the country, as the drug war reduced investor confidence. Deal volume fell below $1 billion per year until 2011.

Civil order largely returned, helping to stabilise investor confidence. Just as important, financial reforms introduced the FIBRA structure in 2012 allowing small land holdings that were a relic of the Mexican Revolution to be transferred into REIT-like structures in which other investors could participate. With new investment vehicles for capital, deal volume surged to $6.8 billion in 2013. In the 12 months to Q1'17, deal volume in Mexico is now down to $3.0 billion.

Looking elsewhere, Colombia has made remarkable moves towards civil order with the FARC peace deal and economic growth and property investment may follow. In May 2016, the largest property deal ever in the region occurred in Chile with Inversiones la Construcción purchasing a portfolio of ten malls from Walmart. At the moment, however, Brazil and Mexico still represent 80% of the investment market in the region, so most of the rest of the analysis will focus on these two countries.

The story of the FIBRAs is a case of changing legal structures, which allowed more investment to occur and more assets to flow into the commercial property sector. There had been a legal structure in place limiting the supply of assets in that owners of older, less efficiently scaled holdings were

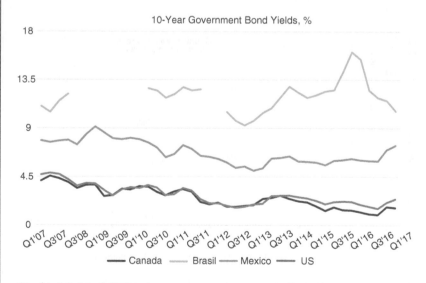

Graphic 6.3 Much higher interest rate environment in Latin American countries analysed

Sources: Real Capital Analytics, Bloomberg

prevented from selling in many cases. Broadly speaking though, the supply of assets does not come as quickly in Latin America as in other regions.

Property investments can look attractive in markets where economic volatility is high as are inflation expectations. It was not too far in the past where Brazil experienced inflation in the range of 1,000% per year. In such an uncertain environment, individuals and companies pursued a number of strategies to protect their wealth.

In the office market for instance, in most parts of the developed world companies tend to lease the space they occupy. In Latin America, however, strata title ownership is common. This structure involves tenants owning the space they occupy in a building. In this way, a business tenant in Brazil, for instance, was protected from rent increases when inflation was raging in the 1970s and 1980s.

Old habits are hard to break, however, and property ownership remains fragmented. This fragmentation even though inflation in both Brazil and Mexico is far lower than it had been in previous decades, is still high relative to their North American cousins in Canada and the US. Into Q1'17 the yield on a benchmark ten-year government bond in Brazil averaged 10.6% in Brazil versus 1.3% in Canada. Higher inflation expectations drive these higher bond rates, with annual rates of inflation of 9% in Brazil in 2016 versus 1.6% for Canada.

These different rates of return in the bond market and different inflation expectations have significant implications for the property investment markets in Latin America. If, for instance, either Mexico or Brazil were able to tame expectations of high rates of inflation, it would unlock significant wealth tied up in commercial property.

As shown in the chart in Graphic 6.4, high real interest rates in Brazil and Mexico are matched by higher average commercial property cap rates relative to their North American counterparts. These two Latin American economies saw an average 8.6% commercial property cap rate in 2016 on top of a 3.2% real interest rate. By contrast, commercial property cap rates stood at 6.2% on average in North America with the real interest rate at 0.1%.

Consider a scenario where the owner of a strata title asset in one of these Latin American markets receives an appraisal stating that their interest in a property is valued at USD 7.5 million. They intend to hold this asset for some time but in this period, suppose a series of policy reforms tames inflation expectations and brings interest rates and cap rates in line with the markets of North America. This investor, simply by holding the asset, would see the valuation rise just above $10 million.[7]

Such reforms could alter the tenure choice of corporate space users and lead to more unitary ownership of assets as is seen in the office markets of developed economies. Still, the whole market in these countries is not held in strata-title arrangements.

There are a number of highly valued assets held by investors as single entities. Given the economic features discussed such as still-high inflation

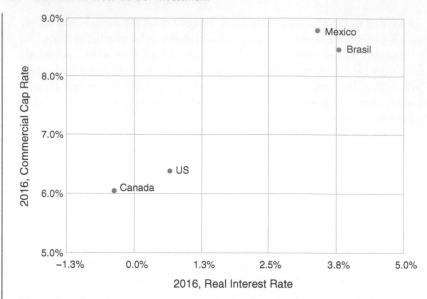

Graphic 6.4 Switching positions in this chart would unlock significant wealth in Latin America

Sources: Real Capital Analytics, Bloomberg; IMF World Economic Outlook, October 2016

expectations and some uncertainty, these investors too are less likely to sell assets than their counterparts in developed economies. Hard assets are a natural hedge against economic uncertainty and these investors are less likely to sell these jewels of their portfolios without an ability to replace these assets without securing a high-quality asset elsewhere.

Motivations of cross-border investors

Still, despite the relative illiquidity of the Latin American markets, some cross-border investors persevere and find ways to put capital to work in the Latin American markets. The origins of the investors able to do these deals and the property types in which they invest both help to explain the motivations of the cross-border actors that are active in the region.

Over the last ten years, cultural and geographic connections have been clear motivating forces driving cross-border investment into Latin America. Investors from the US and Canada have accounted for 54% of all cross-border deals in this time frame, a share that has grown to 56% over the last five years.

Why should US and Canadian investors seek out opportunities in Latin America? There certainly are a number of investment opportunities in North America, but again, the yields are much lower. Canada in particular is a high-income economy with a population base comparable to the size

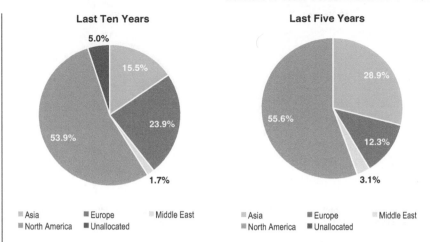

Graphic 6.5 Origin of cross-border capital invested in Latin America . . . mostly
from the North

Source: Real Capital Analytics

of Peru. Over the last decade as petro wealth from the tar sands in Alberta
built up, that wealth needed to be invested somewhere and there are only
a limited number of high-quality assets in which Canadian investors could
put that capital to work domestically.

This high wealth–limited asset relationship drove Canadian investment
in Latin America up to $4.5 billion over the period from 2012 to 2016.
This graphic is $1 billion less than the investment by the far larger pool of
US commercial real estate investors. The ability to put capital to work in
comparatively high yield investments is attractive to these investors.

European investors have been the next largest group of cross-border
investors in Latin America over the last ten years. This investment was led
by German, Swiss and Spanish investors. Like the Canadians, the Germans and Swiss had an excess of wealth and few higher yield opportunities
in their domestic markets. Spanish investors, by contrast, have many more
high-yield opportunities in their domestic markets, but long standing cultural connections make investment easier in this region.

Over the last five years though, investors from Asia have overtaken the
European investors as the second leading group of investors in the region.
This Asian group though is largely Singapore, China and a limited amount
of South Korean investment. The motivations and activities of all these
groups vary.

Singapore is perhaps the ultimate example of a small, wealthy country, and commercial real estate investment must often be focused on cross
border activity. Much of the deal volume from Singapore involved the

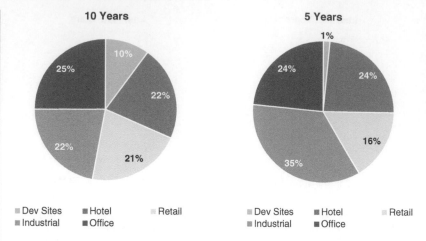

Graphic 6.6 Cross-border investors have pulled back on development recently

Source: Real Capital Analytics

partnership of the country's sovereign wealth fund, GIC with Global Logistics Properties, GLP, a publicly traded investment vehicle focused on the logistics sector.

These Singapore investors and many other cross-border investors from North America have focused heavily on industrial activity of late, accounting for 35% of all deal volume over the last five years. This industrial logistics story is driven by a two-part view of the nature of the Latin American economies within the global economy.

One view is that large population bases in Brazil and Mexico contain domestically a large retail consumption potential in the future as economic liberalization marches onward and helps boost the middle classes in these countries. The thinking is that investments in logistics activities today have both a current high yield and a significant upside potential.

Another view is that the move to offshore US manufacturing activity is becoming a move to nearshore that activity to countries where transit costs and tariff barrier are lower. Mexico in particular has benefitted from such activity given the NAFTA ties to the US. While the new administration dithers on the subject of NAFTA, for now at least the economic ties are in place and certainly transportation costs to this lower-cost manufacturing region remain low.

The hotel sector is also heavily represented in the activity of cross-border investors. The share has been remarkably consistent over both the last five and ten years at 24% and 22% of all transactions, respectively. Much of this activity is focused on the resort like properties in the Caribbean in those small countries with economies tied to the tourism activity from North America and Europe.

Investing in Latin America

Talking with real estate professionals active in Latin America, they will typically try to downplay expectations regarding the ability to execute investment transactions in the region. The ownership is too disaggregated or in the case of the best-of-the-best assets, too concentrated in the hands of investors with no interest in selling.

Latin America, as a destination for capital, represented only 1% of cross-border activity worldwide since 2012; this despite an economy that is 7% of global activity. And yet these real estate professionals and some cross-border investors persevere and find transactions in the region.

One of the main attractions for the region is the high yield opportunity. With low cap rates in the developed economies, the extra spread on offer can look attractive to cross-border investors. It is not free money, though; that higher yield on offer comes with risks from economic and political volatility.

Besides yield though, the region offers scale for investors from wealthy countries with few investible assets. That and to providing space for economies that should have burgeoning middle classes.

6.1 Barriers to FDI for real estate

Some countries try to eliminate or lessen the impact of the barriers that are most likely to isolate the local market from the global capital market. These barriers have been classified by academic work as formal and informal or direct and indirect barriers (Baum and Murray, 2010, see also Eichengreen, 2001 for an overview of more general barriers to investment options that are not exclusive to real estate). The formal or direct barriers are those that primarily affect the ability of foreign investors to invest in emerging markets, for example in the form of taxes and laws. The informal or indirect barriers are those that affect investor's willingness to invest, mainly due to reservations regarding cultural or political issues. Barriers to investment between the parent and host country will be different depending on the type of investment contemplated. For example, tax incentives that a multinational receives for relocating its manufacturing plant to a host country have been known to be more substantial than those received by an insurance company investing in commercial property in the same country (Lahiri, 2009). On the other hand, other costs such as transport and the level of skills of the working population are not likely to be a major barrier to real estate investment, but will be a deterrent for producers. In an investment context and for real estate, Baum and Murray (2010) present the view that formal barriers are known variables that will affect either the ability to invest or the net return delivered; informal barriers represent risks, which may affect the ability to invest or the net return delivered.

Following these authors, the formal and informal barriers to real estate investment can be further classified for real estate as follows:

Formal barriers: a) restrictions to capital accounts; b) legal barriers; and c)
taxation and transaction costs.
Informal barriers: a) political risk; b) market transparency; c) economic stability, currency and liquidity risk; and d) cultural and geographical barriers.

The next sections of this chapter will consider these barriers in the South American context, analysing each one according to GRETI and EFI indices. The GRETI
measures transparency of real estate markets. The index components are: i) performance measurement, which assesses the performance of listed vehicles and
tracks private real estate fund indices; ii) market fundamentals, which measures
the availability of market data for all asset classes including commercial, industrial and residential; iii) governance of listed vehicles, which scores the level of
corporate governance and financial disclosure in countries; iv) regulatory and
legal, which measures property registration, building controls and enforceability of contracts; and v) transaction process, which measures bidding process and
professional standards of agents. According to this index, countries are rated as:
a) highly transparent; b) transparent; c) semi-transparent; d) low transparency;
or e) opaque. Also, the index suggests that the US and the UK are highly transparent while Argentina, Brazil, Chile and Peru are semi-transparent. Colombia,
Ecuador and Uruguay are low transparency; Venezuela is opaque while Bolivia
and Paraguay are not considered by GRETI.

The EFI is issued yearly and it measures the freedom of choice individuals have
in a country in acquiring and using economic goods and resources. Data for the
2017 index covers the period from mid-year 2015 to mid-year 2016. The index is
based on 12 quantitative and qualitative factors that are grouped in the following
categories: a) rule of law (property rights, government integrity, judicial effectiveness); b) government size (spending, tax burden, fiscal health); c) regulatory
efficiency (business, labour, monetary freedom); and d) open markets (trade freedom, investment freedom, financial freedom). Research by the Heritage Foundation shows that economic freedom is highly correlated with higher levels of GDP
per capita; also, countries with more economic freedom tend to have democratic
governments and show signs of faster social progress (2017). The highest score for
this index is 100, which represent the most economic-free countries, although no
country ever reached that level and the best scores in the index are those ranking
above the 80 mark. Hong Kong is the most economic-free nation in the world
while Argentina ranks at the bottom due to government intervention in the
financial sector in past years. Chile, on the other hand, ranks highest in South
America and performs even better than the US.

Given that both indices have been partly built using qualitative data provided
by surveys of a range of actors in relevant sectors, and particularly the real estate
sector in the case of GRETI, the combination of both can help to provide an illustration of the perception potential FDI investors are likely to have of the region.

6.1.1 Formal barriers

According to Baum and Murray (2010), these barriers include restrictions to the investor's ability to remove invested capital from the host country as well as legal barriers that relate to the foreign ownership of local assets. In this category the authors include:

i) Restrictions to capital accounts

This is one of the most common restrictions host countries implement when trying to control the so-called "push and pull" factors. These are terms used in economics to explain international capital flows. Push factors can be related to the lack of lending available in the investors' country, while pull factors are related to the risk-return relationship in the host country (Montiel and Reinhart, 1999). Therefore, push factors explain external reasons why investors choose to go abroad or not; on the other hand, pull factors include some counter-cyclical policies that some countries apply when faced with a surge in the inflow of capital.

One of the components of the EFI scores the Investment Freedom of countries, by measuring the restrictions on the movement of capital, both domestic and international. Table 6.1 shows the ranking of countries under study in this volume according to this component. It also presents the UK and US for comparison. The higher the score, the less restrictions to capital accounts that a country has.

As seen in Table 6.1, the most restrictive countries are Bolivia, Ecuador and Venezuela. The least restrictive are Chile, Colombia and Uruguay, which rank close to the UK and the US. These are followed by Paraguay and Peru, while

Table 6.1 Investment freedom ranking

Country Name	Investment freedom level
UK	90
Chile	85
Uruguay	85
Colombia	80
US	80
Paraguay	75
Peru	75
Argentina	50
Brazil	50
Ecuador	35
Bolivia	5
Venezuela	0

Source: Global Competitiveness Index 2016–17

Data: Investment freedom

Argentina and Brazil apply a certain degree of capital controls to FDI, which in the global context is classified as restrictive (Heritage Foundation, 2017).

The scenario has not always been the same for the Big Five in South America under study here. Table 6.2 shows historical data from the EFI. The series have been compiled according to changes in government administration occurring in each country (marked in the table by the shaded cells). The table shows that Chile and Brazil have been consistent with their levels of investment freedom since 1995, regardless of government change. The difference between the two countries is that Chile has always performed as the less-restrictive country within the Big Five, while Brazil has been at the other side of the spectrum. Colombia and Peru have had two short spells of applying tighter controls to FDI. In the case of Colombia, higher restrictions were introduced under President Álvaro Uribe's term (2002–10) and his own administration gradually eased this policy. In the case of Peru, the higher restrictions coincide with the presidency of Alan García (2006–2011) who briefly imposed restrictions in 2006, but his own administration eased these controls, achieving a score of 60 by 2008 and 2009.

Table 6.2 shows that political changes are not likely to affect decisions on capital controls in the Big Five, with the clear exception of Argentina. This country adopted the liberalisation policies recommended by the World Bank during the 1990s (as did all other countries with the exception of Brazil). These liberal policies were consistently applied until the advent of the new government of Néstor Kirchner (2003–2007), which attempted to control the economic crisis that affected the country from 1998 until 2002 by a protectionist strategy: an approach that was followed by his successor even when the country's recession appeared to be easing around 2009–2011. The current government of President Macri (2015–) has clearly relaxed restrictions from the previous administration, but it remains to be seen if policies will continue to ease and follow Chile's policies, or if they will take the more cautious path presented by Brazil. (At this point the reader might choose to go back to Expert box 5.4 to read a criticism of liberal policies applied in Chile during the 1990s and their effect today.)

ii) Legal barriers

This barrier relates to ownerships restrictions, general rule of law and enforcement of real estate contracts (Baum and Murray, 2010). The EFI includes two variables that can illustrate the level of legal barriers for all countries under study here. These variables include property rights, which scores the security of property (both intellectual and real estate) and judicial effectiveness, which scores countries' ability to enforce the rights of citizens. Table 6.3 shows the ranking of all countries according to the 2017 Index. In the case of property rights, the low scoring of most countries in the index is mainly related to enforcement of intellectual property and data protections issues rather than real estate issues. Bolivia and Paraguay are, however, the exception as, according to the Heritage Foundation (2017), land titling issues remain a high barrier for real estate operations of any kind in Bolivia, while lack of cadastral information makes difficult the registration of property in Paraguay.

Table 6.2 Investment freedom and government change

Country name	2017	2016	2015	2014	2011	2010	2007	2006	2003	2002	2001	2000	1999	1998	1995
Argentina	50	30	30	30	45	45	50	50	50	70	70	70	70	70	50
Brazil	50	55	50	55	50	45	50	50	50	50	50	50	50	50	50
Chile	85	85	90	90	80	80	70	70	70	70	70	70	70	70	70
Colombia	80	80	80	75	65	55	50	50	70	70	70	70	70	70	70
Peru	75	70	70	70	70	70	50	50	70	70	70	70	70	70	70

Source: Economic Freedom Index, www.heritage.org/index/about

Data: Authors' own elaboration

Table 6.3 EFI property rights and judicial effectiveness ranking

Country name	Property rights	Judical effectiveness
Argentina	32.4	39.6
Bolivia	25.7	15.4
Brazil	55.0	49.7
Chile	68.2	63.7
Colombia	63.8	25.2
Ecuador	38.7	22.3
Paraguay	38.2	23.3
Peru	58.3	28.2
Uruguay	70.2	66.8
Venezuela	6.8	10.3
UK	93.8	93.0
US	81.3	75.1

Source: Economic Freedom Index, www.heritage.org/index/about

Data: Author's own elaboration

Judicial effectiveness shows that the best performing country is Uruguay while the least effective is Venezuela. Chile has 63.7 points and, although this is very near the 66.8 scoring for Uruguay, they are both still behind the best international levels – for example, that of the UK (93.8), which has one of the best judicial efficiency levels around the world. According to the Heritage Foundation (2007), Brazilian courts are overwhelmed with current corruption scandals involving most of the political elite. Colombia and Peru's judicial systems are less efficient, with reports of corruption and extortion related to drug trafficking. On the more positive side, the Constitutional and Supreme Courts in Colombia are showing a degree of independence from the executive in recent years, which is helping the country to improve the judiciary efficiency. The same can be said for Argentina that since the departure of the previous administration (which strongly controlled the Courts) new and independent judges have been appointed to the Supreme Courts; this certainly is already a sign of a turn towards a more independent judiciary. Ecuador's justice system is very weak with reports of bribery in exchange for favourable decisions and faster resolution of legal cases. Paraguay presents a similar scenario of corruption in the courts, with cases taking a long time to resolve or being at the mercy of the political elite.

The results provided by the EFI can be cross-examined with the GRETI's regulatory and legal process component, which, as stated before, measures property registration, building controls and enforceability of contracts and is therefore more specific to real estate.

As Table 6.4 shows, Uruguay is the best performing country while Venezuela is the worst. This is consistent with the results shown in Table 6.3 with EFI data. Not surprisingly, therefore, GRETI does not have available information for Bolivia and Paraguay which the EFI also considers highly unreliable in terms of

Table 6.4 Regulatory and legal GRETI index

Country name	Regulatory and legal
Argentina	2.4
Bolivia	n/a
Brazil Tier 1	2.3
Brazil Tier 2	2.3
Chile	2.4
Colombia	3
Ecuador	2.9
Paraguay	n/a
Peru	2.7
Uruguay	1.9
Venezuela	4.6
UK	1.3
US	1.4

Source: JLL global real estate transparency index

rights on real property. Colombia, Ecuador and Peru stand out as having poor regulations and legal frameworks to enforce contracts, while Argentina, Brazil and Chile are in a better position. Notwithstanding, it is clear from both EFI and GRETI indices that all countries are far below the standards of the Anglo-Saxon nations, which, as stated before, performed better for attracting international real estate investment.

iii) Taxation and transactions costs

As it is clear from Expert box 6.1, key drivers for buying global property in detriment of the home country are the increased opportunities for diversification and enhanced return. If returns in the home country are low, investors will seek other alternatives but the increased costs of execution of the operation must be taken into account even if they can be compensated by the potential high returns of the operation.

As seen in the previous section, most countries are below the global average in terms of property rights and judicial efficiency. In contrast, as shown in Table 6.5, some of them score highly in the EFI for taxation and incentives to investors.

Indeed the table shows that Colombia, Ecuador, Paraguay and Peru have less of a tax burden than the US and the UK. There are arguments that these kinds of incentives can be seen as economic weakness as countries with low productivity are the ones that lower the barriers to attract more FDI (Loungani and Razin, 2001). Certainly, in the case of the South American nations, this economic freedom comes at a cost, as these are also the nations that have a greater degree of informality (on average around 50% of the population; see Table 1.10). Not surprisingly, the GRETI component that relates to transaction costs for real estate operations is higher in these nations (see Table 6.5). As stated in Chapter 1, the

Table 6.5 Taxation and transaction costs

Source	EFI	GRETI
Country name	Tax burden	Transaction process
Argentina	62.6	2.3
Bolivia	86.1	n/a
Brazil Tier 1	70.1	2.2
Brazil Tier 2	n/a	2.5
Chile	77.6	2.9
Colombia	80.1	3.3
Ecuador	79.1	3.4
Paraguay	96.2	n/a
Peru	80.3	2.6
Uruguay	65.1	3.2
Venezuela	65.3	4.3
UK	77.5	1.3
US	72.5	1.3

Data: Authors' own elaboration

informal economy is a real burden to South American countries, and governments tend to compensate the lack of personal tax revenues by imposing higher transaction costs to operations that can be easily traced by the governments (see, for example, the imposition of the Unidades Indexadas explained in Chapter 4 as a mechanism to reduce the informal economy in Uruguay).

Most of the countries under study here are part of multilateral tax treatises either within the framework of the Mercosur or the CAM agreements. They also have separate bilateral cooperation agreements that might offer tax incentives to other countries. Peru, for example, has double tax treatises with some CAM and Mercosur countries (Brazil, Chile and Mexico), but also from other regions including Canada, South Korea, Portugal and Switzerland. These tax treatises generally follow the OECD model and limit the withholding of tax rates to 15% of the income generated in the country (Avendaño and Puiggros, 2016). Argentina offers tax incentives to foreign lenders but any non-Argentine with a permanent residency in the country must pay tax on their Argentine-source income.

Some countries are also beginning to offer incentives to foreign investors, particularly when this is related to infrastructure. Brazil, for example, developed a system of infrastructure debentures that have 0% rate of the withholding income tax (provided certain requirements are met). Chile has a reduced withholding tax rate of 4% for interests paid to foreign financial institutions as long as they comply with the requirements of the Chilean tax legislation. Interest payments to non-Chilean nationals with residency in the country can also benefit from a reduced withholding tax rate. Peru also has a reduced withholding income tax rate of 4.99%.

In terms of costs, mortgage fees vary, but they usually involve a percentage of the amount being secured by the collateral limited to a cap. There are also other charges, including legal and land registration fees which can vary from country to country. As an example Argentina has the following charges:

Legal fees: 1% of the secured amount
Stamp tax: between 1% and 1.2% of the secured amount depending on the jurisdiction
Registration fees: 0.2% to 0.3% of the secured amount depending on the jurisdiction

Peru uses a progressive system for land registration fees in relation to the secured amount of the mortgage, being 0.75/1,000 for amounts less or equal to USD 10,000 and 1.5/1,000 for amounts exceeding that limit.

In terms of time for mortgage deeds registration, this varies greatly depending on the assets involved, the size of registry (small towns might take longer) and the workload at the time of registration. As a general reference, in Argentina the registration can take one to six months, in Chile ten to 40 business days, while in Peru the government has set up a fixed period for land registries to complete the operation in less than 60 business days (see Moreno and Chighizola, 2016, for Argentina; Simões Ruso and Rodriguez Cruz, 2016, for Brazil; Peralta and Yubero, 2016, for Chile; and Avendaño and Puiggros, 2016, for Peru).

6.1.2 Informal barriers

Informal barriers to international investment arise because of differences in available information, accounting standards and investor protection. There are also risks that are especially important in emerging markets such as currency risk, political risk, liquidity risk, economic policy risk and macro-economic instability (Bekaert, 1995; Nishiotis, 2004).

According to Baum and Murray (2010) informal barriers for real estate investment include:

i) Political risk

For international investors, this can be a low barrier if the country has a strong economy and an acceptable legal framework. As South America's economy is experiencing slow growth (see Chapter 1) and as demonstrated above the legal framework in most nations is weak; political risk can be an additional important deterrent to FDI operations in real estate in the region.

Table 6.6 shows scoring for the government integrity component of the EFI for the Big Five countries. Following the method used in Table 6.2, which shows government change by cells shaded in grey, the series also highlight the changes in administration for each country via the shaded cells.

As seen in Table 6.6, Colombia and Peru have made slow but steady progress since the beginning of the series in 1995. Both countries have suffered internal conflicts headed by militant groups such as the Shining Path in Peru and the

Table 6.6 Government integrity and change of administration

Country name	2017	2016	2015	2014	2011	2010	2007	2006	2003	2002	2001	2000	1999	1998	1995
Argentina	38.2	34	34	29.5	29	29	28	25	35	35	30	30	28	34	50
Brazil	33.4	43	42	37.9	37	35	37	39	40	39	41	40	36	30	50
Chile	70.5	73	71	72.3	67	69	73	74	75	74	69	68	61	68	50
Colombia	39.6	37	36	33.2	37	38	40	38	38	32	29	22	22	27	10
Peru	38.8	38	38	34	37	36	35	35	41	44	46	45	30	30	10

Source: EFI, www.heritage.org/index/about

Data: Authors' own elaboration

FARC in Colombia. Although both movements appear to be now under control, their activities in the 1980s and early 1990s are mostly responsible for the low scoring received at the start of the EFI series. However, progress has not been as steady as it might be hoped by investors. For instance, in Peru, political corruption is rife with past president Alberto Fujimori currently serving time for bribery and human rights violations; while another past leader, Alejandro Toledo, is wanted by Interpol for allegedly taking bribes from Brazilian construction company Odebrecht (see Expert box 5.1). Nevertheless, Peru has achieved its highest government integrity score during the time of President Toledo (2001–2006), possibly due to his liberal policies and market-friendly policies.

As for the other countries, Argentina's past president was charged in December 2016 over allegations of corruption in infrastructure projects, while most of Brazil's political class is under scrutiny following the impeachment of the previous head of state Dilma Rouseff (2011–16). Even the more stable Chile is facing rumours of nepotism connected to family members of the incumbent Michelle Bachelet. The countries are a long way away from the US and the UK, which (even with their ongoing issues such as Brexit and impeachment rumours of US President Donald Trump) score 78.3 and 78.1, respectively, for government integrity.

ii) Market transparency

Investors and fund managers typically allocate capital to regions and countries before selecting buildings or funds (Baum, 2009). The main arguments for country relevance are to do with the way data, for example national accounts as well as important real estate market data, is collected and made available to the public. Having an independent national data gathering agency and a robust framework for data collection and sharing, can make a significant difference to market transparency. For example, and as shown throughout this volume, Argentina's national data collecting agency (INDEC) was severely reprimanded by the IMF and other international organisations for not publishing accurate information. The result of the dispute was settled when the current administration initiated a new data series and recommended that all data from the country prior to 2015 should be disregarded by the international agencies (see endnote 3 in Chapter 1).

The GRETI collects information regarding availability of data for market fundamentals in all sectors: commercial, residential, and industrial. Table 6.7 presents the scoring for each country under study here as well as UK and US for comparison.

Table 6.7 shows that, with the exception of tier 1 cities in Brazil, the rest of the countries have scant market fundamentals information available for investors.

Furthermore, some of the data available for the residential sector does not record the sale price of units, but the valuation estimate given by the agencies. The property market is characterised by its heterogeneity; this means individual properties have their own characteristics and they are not all the same. Even two similar properties in the same block of houses can be valued differently

Table 6.7 GRETI market fundamentals

Country name	Market fundamentals	Transparency level
Argentina	3.7	semi-transparent
Bolivia	n/a	n/a
Brazil Tier 1	2.9	semi-transparent
Brazil Tier 2	3.3	semi-transparent
Chile	4.5	semi-transparent
Colombia	4.3	low-transparency
Ecuador	4.4	low-transparency
Paraguay	n/a	n/a
Peru	4.3	semi-transparent
Uruguay	4.5	low-transparency
Venezuela	4.5	opaque
UK	1.5	high-transparency
US	1.4	high-transparency

Notes: GRETI divides Brazil into two levels of cities according to the size of the market. In the case of Regulatory and Legal both tiers score the same

Source: JLL global real estate transparency index

because one has been modernised and viewed by the valuer as more expensive than the other. Valuers will advise on comparable properties and provide a best estimate, but the final price is the purchase price, which records the true value of the asset in the market. The problem in most South American countries is the lack of information on completed transactions rather than valuations, which is the most common source of information available in open sources. In recent years, some companies such as Reporte Inmobiliario in Argentina have started a database for residential properties but this is only based on valuations and not transaction prices. Even so, this knowledge varies geographically and not all locations are as well reported as capital cities. The example of the difference in market transparency between tier 1 and tier 2 cities in Brazil, is a testimony of this geographical disparity in the level of information. In addition, the knowledge in most countries tends to be infrequent and lacks up-to-date valuations.

Another problem is presented by incentives as they can generate market distortions. In a "perfect market," consumers and producers seek to maximise profits and utility in an environment unhampered by legal or other constraints: for example, perfect knowledge of the market, no ostentatious buying and no advertising that can influence consumer's behaviour. However, these static assumptions are impossible in real life (Harvey and Jowsey, 2004). As seen in Chapters 3 and 4 for the residential market and investment, governments can and do affect the supply of housing by providing tax credits to incentivise development. Equally, governments offer subsidies to potential buyers to facilitate their access to mortgage market and increase demand, such as the help-to-buy schemes in the UK. Understanding these incentives is important for investors who might otherwise interpret a high demand as a healthy market in need of more development, while in fact, it is mainly down to government incentives. Once the incentives are reduced as a result of policy changes, the demand reduces considerably, potentially leaving the market with an oversupply of new units. A similar scenario occurs in commercial property, for example, when leases are signed with a rent-free period to entice potential tenants. Sometimes these are disclosed as a "fit-out" clause, which usually involves a period of time where the prospective tenant pays no rent or a lower rent in order to cover moving costs of the business. In a two-year commercial lease, a rent-free period or fit-out clause for, say, two or three months can considerably alter the average rental price per square metre of the commercial asset and in turn its capital valuation (Crosby and Murdoch, 1994). The non-disclosure of such arrangements tends to distort the markets and can influence the degree of real estate transparency in a country.

iii) *Economic stability, currency risk and liquidity risk*

The economic stability is one of the most important factors that is likely to influence investor's decision-making for real estate FDI (Baum and Murray, 2010). These authors' survey results show that even if the political scenario is fragile, provided that the target country has economic stability and can offer good returns, then all other risks can be costed and added to the operation. However, currency risk can be an important barrier if investors cannot hedge or if the costs of doing so are too high. Academic research shows that for real estate operations, currency hedging is expensive and difficult to achieve efficiently (Lizieri et al., 1998) and vehicles are rarely fully hedged. This problem can leave investors at the mercy of currency fluctuations. In the case of liquidity, the depths of a country's capital market is important as these can guarantee that there will be sufficient investors to trade operations.

Table 6.8 shows the scoring for three index components that can serve as a proxy to measure these informal barriers in all countries under study here. The first column is the EFI variable for Fiscal Health, which measures the average deficit and the overall debt burden per country as percentage of GDP. The second column is the EFI variable for Monetary freedom, which takes into account the monetary policies implemented by countries and in particular if they endeavour to fight inflation and maintain price stability. The last column of the table presents the GRETI component for investment performance, which tracks the performance of several indices including direct property, listed real estate securities and private real estate funds. This serves as a proxy for the health and size of the real estate market and therefore its liquidity. The combination of all these variables will help to illustrate how well equipped these countries are to break down these barriers.

As previously explained, FDI country scoring is higher as perceptions of economic freedom improve; on the contrary, the GRETI's best performers have a low scoring as transparency improves. As seen in Table 6.8, countries with a low debt-to-GDP ratio will score high in the EFI fiscal health column; these include

Table 6.8 Economy stability, currency and liquidity risk

Source	EFI	EFI	GRETI
Country name	Fiscal health	Monetary freedom	Investment performance
Argentina	56.4	50.9	4
Bolivia	81.4	66.4	n/a
Brazil	22.8	67.0	3
Chile	96.1	82.2	3.8
Colombia	89.8	77.0	4.6
Ecuador	56.4	67.7	4.3
Paraguay	95.1	78.3	n/a
Peru	98.4	83.3	4
Uruguay	77.2	71.3	4.7
Venezuela	15.2	16.8	4.5
UK	40.4	85.0	1
US	53.3	80.1	1.2

Data: Authors' own elaboration

Bolivia, Chile, Colombia, Paraguay, Peru and Uruguay. Countries with a medium score are Argentina and Ecuador while Brazil and Venezuela have the lowest scoring. Still, as explained in Chapter 1, there is no agreement amongst economists on a maximum or minimum ratio for the debt-to-GDP relation, and this is more related to the ability of countries to service and honour their obligations. In this sense, these figures should be compared to the rating of countries provided in Table 1.22. According to this table, the best countries in the region are Argentina and Peru, both rated between positive and stable, followed by Colombia, which has only one negative rating by S&P, while Brazil and Chile both have more negative than stable ratings and no positives.

In the case of monetary freedom, the chart shows that Chile, Colombia, Peru and Uruguay have what the EFI considers a good level of monetary freedom. The rest of the countries, particularly Argentina and most noticeably Venezuela, are fighting inflationary trends in consumer prices. The performance of real estate investment markets in all South American nations remains very distant from the Anglo-Saxon countries. Brazil is the only nation showing that it is by far the best regional performer while Uruguay, with its steady government and rule of law, scores even lower than Venezuela, a country that is usually ranked at the bottom of all indices and charts not only at regional but also at global level. The low liquidity level in most countries presents an important barrier to real estate FDI (see also Expert box 6.1). Increasing data availability and transparency is therefore crucial for South America if they want to improve international investment.

iv) Cultural barriers and geographical barriers

As survey results from Baum and Murray (2010) show, these are soft barriers that relate to education of the workforce and the ability to communicate with locals in the English language. Other factors, for example religion, have been disregarded as of minor importance by investors. Geographical barriers relate to regional and country choices for FDI investment. This barrier is even more relevant for real estate, as spatial characteristics are a key feature of this asset class. The interaction facilitated by spatial proximity helps to build the trust and rapport, which is vital as investors gather market information (Leyshon and Thrift, 1997; Agnes, 2000). For these reasons, geography still matters for portfolio choice, savings and investment, and can have a great influence on investors' decisions and returns (Stulz, 2005). Not surprising and as shown in Expert box 6.1, most of the FDI investment into Latin America comes from their northern neighbours, mainly the US and Canada. Other perceived difficulties, including the dangers of operating from a distance with no local representation, increases the attraction of investing internationally through liquid securitised vehicles and unlisted funds, but geography remains a barrier to international exposure by asset managers.

South America as a whole is a region where English as a second language is widely used, mainly given the influence that North American countries have had on the continent. The Education First English Proficiency Index (EFEPI) ranks countries by their English skills by collecting results of a set of English tests completed by adults around the world. Data presented here is for the 2017 edition of the Index, which is based on 950,000 test-takers who completed three different EF English tests in 2015. The EFEPI classifies countries according to the scoring achieved in five categories, including very high, high, moderate, low and very low.

Table 6.9 shows the ranking of countries and the classification by the EFEPI including other countries in Asia and Europe for comparison.

Table 6.9 English proficiency

Country name	Score	Classification
Portugal	59.68	High
Argentina	58.4	High
India	57.3	Moderate
Spain	56.66	Moderate
Uruguay	51.6	Low
China	50.9	Low
Brazil	50.7	Low
Chile	50.1	Low
Peru	49.8	Low
Ecuador	49.1	Low
Colombia	48.4	Very low
Venezuela	46.5	Very low
Bolivia	n/a	n/a
Paraguay	n/a	n/a

Source: Education First, www.ef.co.uk/epi.

Data: Authors' own elaboration

From the South American nations, Argentina has been classified as having a high level of proficiency in English as well as Portugal in Europe. The southern country performs even better than Spain, another member of the European Union where English language is widely used for communication with other member states. The rest of the countries have a low ranking, with Colombia and Venezuela showing a very poor level of English, while there is no available information for Bolivia and Paraguay. Most South American countries rank the same as China and India in Asia, where the EFEPI regional average is 55.94, higher than for the whole of Latin America, which is 50.54. Asia's higher scoring is driven by countries such as Singapore (63.52) and Malaysia (60.70). This is not to suggest that this barrier is high in South America, but the ability of foreign firms to communicate and build rapport with locals can influence the direction of FDI to other regions with higher English skills, given that the industry is dominated by the Anglo-Saxon speaking nations (GRETI, 2017).

In terms of levels of education and availability of skilled workforce, the ILO provides statistics on the percentage of a country's workforce in employment during a determined period of time, desegregating data by level of education achieved using the International Standard Classification of Education (ISCE) as classification method. Table 6.10 shows the statistics for South America with other countries in Europe and Asia for comparison, presenting the percentage of workforce

Table 6.10 Level of education

Country	Age	Education	Source	%	Year
US	15+ (age)	Advanced (aggregate levels)	Labour force survey	72.1	2016
Singapore	15+ (age)	Advanced (aggregate levels)	Labour force survey	51.6	2015
UK	15+ (age)	Advanced (aggregate levels)	Labour force survey	43.1	2016
Spain	15+ (age)	Advanced (aggregate levels)	Labour force survey	42.1	2016
China	15+ (age)	Advanced (aggregate levels)	Labour force survey	32.9	2015
Peru	15+ (age)	Advanced (aggregate levels)	Other household survey	30	2015
Venezuela	15+ (age)	Advanced (aggregate levels)	Labour force survey	29.9	2012
Portugal	15+ (age)	Advanced (aggregate levels)	Labour force survey	26.3	2016
Malaysia	15+ (age)	Advanced (aggregate levels)	Labour force survey	23.2	2015
Uruguay	15+ (age)	Advanced (aggregate levels)	Labour force survey	22.7	2014
Paraguay	15+ (age)	Advanced (aggregate levels)	Other household survey	22.6	2013
Colombia	15+ (age)	Advanced (aggregate levels)	Labour force survey	21.9	2014
Argentina	15+ (age)	Advanced (aggregate levels)	Labour force survey	21.8	2014
Chile	15+ (age)	Advanced (aggregate levels)	Labour force survey	16.3	2016
Ecuador	15+ (age)	Advanced (aggregate levels)	Labour force survey	16	2016
Brazil	15+ (age)	Advanced (aggregate levels)	Labour force survey	14.3	2014

Source: ILO, www.ilo.org/ilostat/faces/oracle/webcenter/portalapp/pagehierarchy/Page27.jspx?subject=
EMP&indicator=EMP_TEMP_SEX_AGE_EDU_NB&datasetCode=A&collectionCode=YI&_
afrLoop=118659099281283&_afrWindowMode=0&_afrWindowId=11uja8lur7_51#!%40%40%3
Findicator%3DEMP_TEMP_SEX_AGE_EDU_NB%26_afrWindowId%3D11uja8lur7_51%26sub
ject%3DEMP%26_afrLoop%3D118659099281283%26datasetCode%3DA%26collectionCode%3
DYI%26_afrWindowMode%3D0%26_adf.ctrl-state%3D11uja8lur7_71

Data: ILO

with an advanced level of education. Some points of concern related to this data are the lack of homogeneity of survey sources, particularly for Peru and Paraguay. Additionally, not all countries have yet presented statistics for 2016; therefore, the table presents the nearest available option, which, in the case of Venezuela, is from seven years ago. Some of this statistical concern might explain the high ranking of these countries in the table. However, the table illustrates the fact that as with the level of English, Asian countries are still performing better than South America, and that Portugal and Spain have a much more skilled workforce than Uruguay, which is the country that has the most reliable data and the best performance in most other indicators for South America.

Other cultural and geographical barriers, such as travel distance and time difference, are barriers that can be encountered by investors in other countries and are therefore not specific to the region. However, the level of English and education can be an additional factor that is hindering FDI to the region. One of the Millennium Development Goals was to achieve universal primary education for all (Goal 2). Most South American countries did extremely well, with some even surpassing the targets. But the region now has a bigger challenge in Goal 4 of the Sustainable Development Goals (SDG), which intends to better the quality of past achievements in education. This could be an opportunity for countries not to be left behind by the Asian Tigers, as it has been the case during the current century. The next and final chapter of this book will look at how the region is responding to the SDGs and the New Urban Agenda, and what policy implications these international agreements have for real estate and development in South America.

Notes

1 In the context of this volume, FDI is a long-term investment made by a non-resident and with control of at least 10% or more of the investment (see Lahiri, 2009).
2 The authors conducted a series of semistructured interviews with fund managers and real estate investors from around the globe, including the US, Europe and Russia, asking them to rate the relevance of FDI barriers to real estate investment to developing nations.
3 www.jll.com/GRETI
4 www.heritage.org/index/
5 Figures for 2016 estimates from IMF World Economic Outlook database. Using the term Latin America as a shorthand for countries in Central America, the Caribbean and South America. Including Mexico as well, despite being classified as part of North America by some analysts due to the NAFTA ties to Canada and the US.
6 Excluding The Bahamas, and Antigua and Barbuda here. Much higher GDP per capita and with very low population skews the chart to the point of not being able to see other countries.
7 This USD 10 million threshold example is not a random example picked to make a case. Globally, Real Capital Analytics tracks all deals priced USD 10 million and over. Thus with high inflation, interest rates and thereby cap rates, some portion of the Latin American market is excluded from this analysis.

Bibliography

Agnes, P. (2000) The 'End of Geography' in Financial Services? Local Embeddedness and Territorialization in the Interest Rate Swaps Industry. *Economic Geography*, 76, pp. 347–366.

Aitken, B. and Harrison, A. (1999) Do Domestic Firms Benefit From Direct Foreign Investment? Evidence From Venezuela. *The American Economic Review*, 89, 3, pp. 605–618.

Avendaño, J.L. and Puiggros, J.M. (2016) Peru. In *Lending and Secure Finance 2016*. International Comparative Legal Guides Publication. Vancouver: McMillan.

Baum, A. (2009) *Commercial Real Estate Investment: A Strategic Approach*. Exeter, UK: EG Books.

Baum, A. and Murray, C. (2010) Understanding the Barriers to Real Estate Investment in Developing Economies. *Working Papers in Real Estate and Planning, 3/11*. Available at http://centaur.reading.ac.uk/26974/.

Bekaert, G. (1995) Market Integration and Investment Barriers in Emerging Equity Markets. *World Bank Economic Review*, 9, pp. 75–107.

Bengoa, M. and Sanchez-Robles, B. (2003) Foreign Direct Investment, Economic Freedom and Growth: New Evidence From Latin America. *European Journal of Political Economy*, 19, 3, pp. 529–545.

Chuang, Y. and Hsu, P. (2004) FDI, Trade and Spillover Efficiency: Evidence From China's Manufacturing Sector. *Applied Economics*, 36, 10, pp. 1103–1115.

Crosby, N. and Murdoch, S. (1994) Capital Valuation Implications of 'Rent-free Periods'. *Journal of Property Valuation and Investment*, 12, 2, p. 64.

De Mello, L. (1999) Foreign Direct Investment Led Growth: Evidence From Time Series and Panel Data. *Oxford Economic Papers*, 51, pp. 133–151.

Dunning, J.H. (1994) *Re-evaluatin the Benefits of Foreign Direct Investment*. Reading: Department of Economics, University of Reading.

EFI. (2017) *Index of Economic Freedom: Trade and Prosperity at Risk*. Available at www.heritage.org/international-economies/report/2017-index-economic-freedom-trade-and-prosperity-risk, accessed on 9th July 2017.

Eichengreen, B. (2001) Capital Account Liberalization: What Do Cross-Country Study Tell Us? *The World Bank Economic Review*, 15, pp. 341–365.

Garrett, G. (2000) The Causes of Globalization. *Comparative Political Studies*, 33, pp. 941–991.

GRETI. (2016) *Global Real Estate Transparency Index 2016*. Available at www.jll.com/greti/Documents/GRETI/Global-Real-Estate-Transparency-Index-2016.pdf

GRETI. (2017) *Global Real Estate Transparency Index*. Available at www.jll.com/GRETI, accessed on 9th July 2017.

Harvey, J. and Jowsey, E. (2004) *Urban Land Economics*. London: Palgrave.

Heritage Foundation (2017) Available at www.heritage.org/index/about.

Imazeki, T. and Gallimore, P. (2010) Domestic and Foreign Bias in Real Estate Mutual Funds. *Journal of Real Estate Research*, 26, pp. 367–390.

IMF. (2016) *Regional Economic Outlook*. Washington, DC: IMF. Available at www.imf.org/~/media/Websites/IMF/imported-flagship-issues/external/pubs/ft/reo/2016/whd/eng/pdf/_wreo0416pdf.ashx

Lahiri, S. (2009) Foreign Direct Investment: An Overview of Issues. *International Review of Economics and Finance*, 18, 1, pp. 1–2.

Lapoza, S. (2007) *The Foreign Direct Investment Property Model: Explaining Foreign Property Demand & Foreign Property Capital Flows in Transitional Economies*. University of Reading Thesis R10036.

Leyshon, A. and Thrift, N. (1997) A Phantom State? The De-traditionalisation of Money, The International Financial System and International Financial Centres. In Leyshon, A. and Thrift, N. (eds.), *Money Space: Geographies of Monetaries Transformations*. London: Routledge.

Lizieri, C., Worzala, E., and Johnson, R. (1998) *To Hedge or Not to Hedge?* London: RICS.

Loungani, P. and Razin, A. (2001) How Beneficial Is Foreign Direct Investment for Developing Countries? *Finance and Development*, 38, 2. Available at www.imf.org/external/pubs/ft/fandd/2001/06/loungani.htm, accessed on 16th June 2017.

Montiel, P. and Reinhart, C. (1999) Do Capital Comtrols and Macroeconomics Policies Influence the Volume and Composition of Capital Flows? Evidence From the 1990s. *Journal of International Money and Finance*, 18, 4, pp. 619–635.

Moreno, J. and Chighizola, D. (2016) Argentina. In *Lending and Secure Finance 2016*, International Comparative Legal Guides Publication. Vancouver: McMillan.

Nishiotis, G. P. (2004) Do Indirect Investment Barriers Contribute to Capital Market Segmentation? *The Journal of Financial and Quantitative Analysis*, 39, pp. 613–630.

Peralta, D. and Yubero, E. (2016) Chile. In *Lending and Secure Finance 2016*. International Comparative Legal Guides Publication. Vancouver: McMillan.

Sassen, S. (2006) The Embeddedness of Electronic Markets: The Case of Global Capital Markets. In Knorr Cetina, K. and Preda, A. (eds.), *The Sociology of Financial Markets*. Oxford: Oxford University Press.

Simões Ruso, R. and Rodriguez Cruz, L.B. (2016) Brazil. In *Lending and Secure Finance 2016*. International Comparative Legal Guides Publication. Vancouver: McMillan.

Stiglitz, J. (2000) Capital Market Liberalisation, Economic Growth and Instability. *World Development*, 28, 6, pp. 1075–1086.

Stulz, R.M. (2005) A Model of International Asset Pricing. *The Journal of Financial Economics*, 9, 4, pp. 383–406.

Talalay, M. (2000) Technology and Globalization: Assessing Patterns of Interaction. In Germain, R. (Ed.), *Globalization and Its Critics*. London: Macmillan Press.

Wylde, C. (2011) State, Society and Markets in Argentina: The Political Economy of Neodesarrollismo under Néstor Kirchner, 2003–2007. *Bulletin of Latin American Research*, 30, 4, pp. 436–452.

Zhang, K. (2001) Does Foreign Direct Investment Promote Economic Growth? Evidence From East Asia and Latin America. *Contemporary Economic Policy*, 19, 2, pp. 175–185.

7 Guiding improved urban development

UN Sustainable Development Goals and the New Urban Agenda – how is South America faring in the world scene?

7.0 Introduction

"Sustainability" is a wide-ranging term, often bandied about with little understanding of its meaning or intent. The original definition of sustainable development is generally agreed to be:

> Development that meets the needs of the present without compromising the ability of future generations to meet their own needs.
>
> Bruntland Report for the World Commission
> on Environment and Development (1992)

This definition basically means that anything development does now should not compromise resources or opportunities in the future. Sustainable practice can be applied to every aspect of human activity: shelter, food production, mobility, use of natural resources as well as political and social systems. When considering the sustainability of real estate proposals, it is necessary to consider a very wide range of factors, many of which require significant adaptive change. The UN Sustainable Development Goals are a benchmark for what the majority of world leaders now believe society needs to work towards.

This final chapter will focus on how South American countries are performing according to the SDGs, and in particular, in those goals that are closely related to real estate development. The next section (7.1) will therefore explain the SDGs and their predecessors, the MDGs, which concluded in 2015. In section 7.2 the New Urban Agenda signed in Quito (Ecuador) in 2016 will be briefly described. Section 7.3 presents intended, nationally determined contributions agreed during the 21st Conference of the Parties in Paris 2015 (COP21) to combat climate change. These sections prepare the reader for a deeper understanding of the issues the world is facing in relation to climate change, and their applicability to South America, paying special attention to the role real estate development plays in the most urbanised continent on earth. Section 7.4 therefore evaluates how the Big Five are performing by following a series of international indicators, while section 7.5 evaluates each of the Big Five countries and their current policies in order to point out best practice as well as shortcomings. The final section and the conclusions to this volume (7.6) present the author's policy recommendations for the region.

7.1 Sustainable Development Goals

In 2016, 17 Sustainable Development Goals (SDGs) of the UN's 2030 Sustainable Development Agenda came into force, having been adopted by the vast majority of world leaders on September 2015 including all countries in South America under study in this volume. They set out principles of good development that will protect people's rights to a decent quality of life and the protection of the planet's environment. These principles are expected to guide government policy, regulatory authorities, funding bodies and developers in the foreseeable future. The SDGs build upon, and now replace, the Millennium Development Goals that were adopted in 2000, which set targets to address poverty, social inequality and environmental sustainability and set out basic human rights for every person to health, education, shelter and security. The difference between the MDG and the SDG is that the former intended to reduce poverty while the latter intends to eradicate it. This presents a bigger challenge for regions like South America with large sectors of the population living in poverty and informality.

SDGs cover broad issues such as ending poverty and hunger (including food security and sustainable agriculture); health and well-being; gender equality; sustainable management of water and sanitation; sustainable energy supplies; resilient infrastructure; reducing inequality within and across borders; ensuring settlements are safe, resilient and sustainable; combating climate change; protecting and restoring the environment; promoting peaceful and inclusive societies with justice for all; and finally, to strengthen the global partnership for sustainable development. The connection between the SDGs and urban development are plain to see; however, Appendix 1 briefly spells out how each relates to real estate development.

7.2 New Urban Agenda (NUA)

The NUA should not be confused with New Urbanism, an urban design movement established in the 1980s in the US, which has advocated human-scaled urban form and traditional walkable neighbourhoods.

At UN Habitat III (global summits held every 20 years) in Quito in 2016, the UN adopted the NUA, which is a framework for urban development, enshrining the principles of the SDGs. The NUA began with the *Quito Declaration on Sustainable Cities and Human Settlements for All*, a ten-point outline of the challenges and opportunities as the world's urban population is expected to double by 2050. Fundamental to the NUA are equal rights and freedoms for all humans to a decent housing and participation in civic systems. While the SDGs are wide-ranging global principles, their implementation is local. There is emphasis on the participation of civil society, which will mean that the development process and decision making will have to change significantly.[1] Bringing about this change in the way things are done will be complex and challenging, and observers are questioning just how this paradigm shift will be brought about. However, the UN sees the NUA as an opportunity to leverage the role of the urban dwellers as drivers of sustainable development (de Paula, 2016). It is hoped that "localising" the SDGs will promote a more "bottom-up" approach to urban development and lead to a strengthening of local and regional governments. This localisation of the SDGs will be both a political process and a technical one. If local governments can be held accountable by citizens (for instance, if they fail to direct local development), then democratic responsibility could become a powerful tool for achieving the SDGs at municipal level. At Habitat III, local and regional governments recognised the positive impact of empowering citizens to be proactive participants in urban development. They expressed the desire for national governments to place the SDGs at the centre of their planning policy, whilst at the same time acknowledging that local ownership of these strategies is paramount for the successful delivery of the NUA. Bringing about this change will challenge existing institutional processes and the relationships between local and central governments.[2]

7.3 Intended Nationally Determined Contributions

At the UN Framework Convention on Climate Change (UNFCCC) Conference of the Parties (COP 21) in Paris in December 2015, nearly 200 UN delegates reached a historical global agreement to address the threat posed by climate change. The Paris Agreement, as it is now known, sets a common path to reduce global greenhouse gas emissions and achieve a zero-carbon, climate-resilient future. The signatory countries pledged to hold the increase in global average temperature to well below 2°C, to pursue efforts to limit the increase to 1.5°C, and to achieve net-zero emissions in the second half of this century. In preparation for the COP21, countries submitted their Intended Nationally Determined Contributions (INDCs), which are now being ratified by most countries by submitting their Nationally Determined Contributions (NDCs). The INDC or NDC outlines individual countries' intentions for reducing emissions, and are built taking into account a country's domestic circumstances and capabilities. This, in some cases, includes how countries will adapt to climate change impacts and what support they will need or will provide to other countries, towards building mitigations strategies and climate resilience. SDG 13 ("Take urgent action to climate change and its impact") incorporates the Paris Agreement in the 2030 targets agreed by UN members states.

7.3.1 South America's INDCs at the country level

Appendix 2 presents the INDCs for all countries under study here with the addition of China and the US for comparison. Graphic 7.1 shows Green House Gas (GHG) emissions including Land Use Change and Forestry for 2012 for

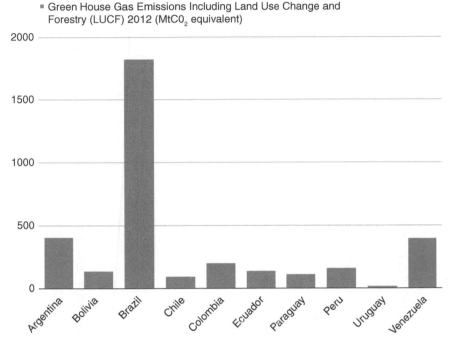

■ Green House Gas Emissions Including Land Use Change and Forestry (LUCF) 2012 (MtCO$_2$ equivalent)

Graphic 7.1 GHG emissions including LUCF

all countries measured in MtCO$_2$ equivalent (MtCO$_2$e). As seen in the figure, Brazil is the highest emitter of the region, with 1,823.15 MtCO$_2$e, while the lowest is Uruguay with 14.76. Argentina and Venezuela are both around the 400 mark (Argentina 405.03 and Venezuela 396.74). Colombia and Peru are over 150 (Colombia is 199), while Bolivia and Ecuador and Paraguay are over 100 but not above 140 MtCO$_2$e.

The top ten world's worst emitters include Brazil as well as the US (5,823 MtCO$_2$e) and China (10,684 MtCO$_2$e). Clearly the rest of the South American countries are not high emitters, but it is policies and future trajectories that matter. Graphic 7.2 presents CO$_2$ emissions per capita for Argentina, Brazil, Chile and Peru (data for Colombia is not available for comparison). Historic data in this graphic already shows an upward trend in CO$_2$ emissions for all countries and future projections if no changes are made to current policies, are also growing, particularly in countries with the highest GDP (Brazil and Argentina).

Additionally, building emissions intensity per capita are also on an upward trajectory in developing countries (Graphic 7.3), while emissions are decreasing in more developed economies such as the UK and US.

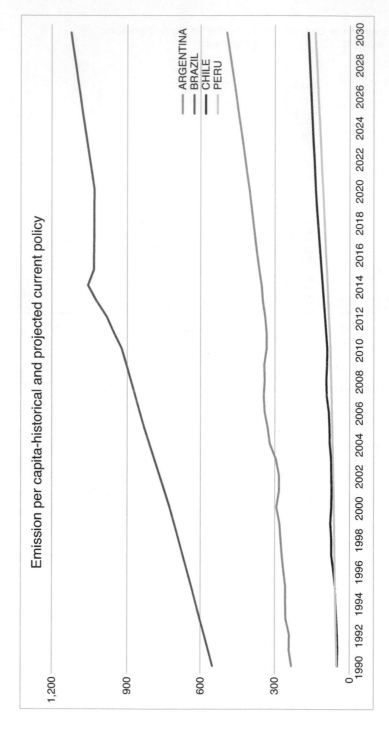

Graphic 7.2 Emissions per capita, historical and projected

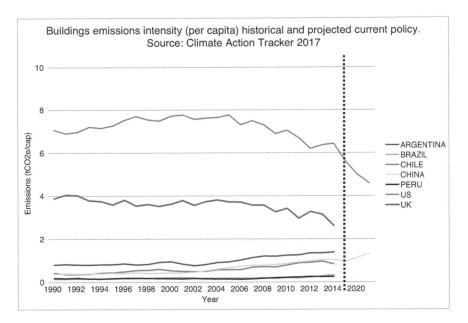

Graphic 7.3 Building emissions intensity

7.4 How South America is performing in the global context

The UK-based Overseas Development Institute (ODI) report *Projecting Progress –
The SDGs and Latin America and the Caribbean*[3] provides an assessment of the prospects for the region. ODI suggests that meeting the SDG targets in Latin America
will require major new efforts. It points out that though progress has already been
made in many areas, some trajectories are changing (e.g. GDP per capita growth
has declined from 3.2% during 2004–2008 to 1.5% in 2014). This has translated to
weaker social gains, lower income growth and slower poverty reduction. However,
the ODI report highlights that most countries in Latin America are set to achieve
the inequality target, though more effort will be needed to end extreme poverty.
The region is also scoring well on the energy targets, with all countries set to reach
over 95% of their populations with access to electricity. Notwithstanding, sanitation and biodiversity and halting deforestation are progressing too slowly to meet
the 2030 target. Goals related to hunger, health, education, gender and partnership are unlikely to meet 2030 targets unless progress rates are increased by five to
eight times the current. Targets related to cities, waste, climate, oceans and peace
are actually moving in the wrong direction and need to be reversed.

There is one goal, SDG 16 on peace and ending violent deaths, where the
region is going in completely the wrong direction, if compared to global averages.
Also, SDG 17 (strengthening implementation) shows that reversals in trends are
needed for South America.

The statistical arm of the UN (UN stats) has grouped available data for each of the 17 SDGs. It is important to highlight that an agreed methodology for tracking SDGs is still being developed by the UN's Inter-Agency and Expert Advisory Group (IAEG). Until that methodology is available, and in order to have some interim process for tracking progress, the SDG Index and Dashboard (SDGID) was launched in July 2016 by the Sustainable Development Solutions Network (a UN global initiative) and the Bertelsmann Stiftung, a philanthropic privately operated organisation. The SDGID is not an official product endorsed by the UN or any government, but provides a temporary and reliable source until an official methodology is agreed by the IAEG.

The SDGID tracks country performance across all 17 goals in order to achieve targets by 2030. A country's position is indicated on a 0-to-100 spectrum from the "worst" (score 0) to the "best" (score 100).[4] The Index tracks 149 countries of the 193 UN members states and uses relevant data already available at UN stats, but filling data gaps with variables published by sources considered reputable by the index provider. The SDGID also present an SDG Dashboard, with each goal coloured as "green," "yellow," or "red," indicating whether the country has already achieved the goal (green), is in a "caution lane" (yellow), or is seriously far from achievement as of 2015 (red). Appendix 3 shows the dashboard for Latin America as a whole; still, what matters most for the purpose of this book is how countries are performing in SDGs that are closely related to urban development.

Looking at all variables to track SDG progress, the most important indicators for real estate development provided by the SDGID are primarily those related to goal 11, *Make cities and human settlements inclusive, safe, resilient and sustainable*. The SDGID variables included for tracking progress in this goal are: pollution measurement in urban areas (measured as PM >2.5 μg/m3 of air), and improved water source piped (measured as a percentage of total). Another important goal is 13, *Take urgent action to combat climate change and its impacts*. Indicators for this goal include: CO_2 emissions from energy (measured as t/CO_2/capita) and climate change vulnerability (estimated in a scale of 0–1).

Additionally, the following list of variables and measurement methods available for South American countries, are also relevant to real estate development:

- SDG 3: Traffic death (as per 100,000 inhabitants);
- SDG 6: accessed to improved sanitation (%); freshwater withdrawal (%);
- SDG7: access to electricity (%); access to non-solid fuels (%); CO_2 from fuels and electricity ($MtCO_2$/TWh);
- SDG 9: quality of overall infrastructure (measured on a scale of 1–7);
- SDG 12: wastewater treated (%); municipal solid waste (kg/person/year); and
- SDG13: CO_2 emissions from energy (t/CO_2/capita); climate change vulnerability (0–1)

Table 7.1 shows the SDGID dashboard for the list of variables mentioned here for the Big Five countries. Amongst the most positive outlooks, is the availability of fresh water sources that the region has, which in the chart is indicated by the

Table 7.1 SDGID dashboard scoring for selected variables

Country Name	Traffic death	Access fresh water	Improved sanitation	Improved water	CO₂ from fuels and electricity	Access to electricity	Access to non-solid fuels	Infrastructure quality	Municipal solid waste	Wastewater treated	Air pollution	CCH vulnerability	CO₂ emissions/ capita
Argentina	red	green	red	red	red	red	red	n/a	green	red	green	yellow	green
Brazil	red	green	yellow	green	green	green	green	red	yellow	red	yellow	yellow	yellow
Chile	red	green	red	red	n/a	red	red	red	green	red	yellow	yellow	green
Colombia	yellow	green	yellow	yellow	yellow	yellow	green	yellow	green	red	yellow	yellow	green
Peru	yellow	green	yellow	yellow	yellow	yellow	yellow	yellow	green	yellow	yellow	yellow	green

Source: www.sdgindex.org

Data: Assembled by authors directly from SDGID datasets

green rating in the second column. The other positives for most countries (with the exception of Brazil) are the low levels of municipal solid waste and low levels of CO_2 emissions/capita (columns 9 and 13). On the other side of the spectrum, countries' waste water management (column 10), death toll by traffic accidents (column 1) and quality of infrastructure (column 8) can be barriers for the Big Five to achieve their SDGs by 2030. Argentina and Chile are perhaps the more worrisome as they show the highest number of red scoring across all indicators. Brazil, on the other hand, is the one with the highest green scoring while Colombia and Peru are remarkably similar sharing both a "cautious lane" yellow path as of 2015 measures for most variables.

7.5 Are policies in the Big Five sufficient?

Climate Action Tracker (CAT)[5] is an independent scientific analysis produced by three research organisations: Climate Analytics, New Climate Institute and Ecofys. Since 2009, CAT has been tracking climate action progress towards the globally agreed aim of holding warming well below 2°C, and pursuing efforts to limit warming to 1.5°C.

CAT ranks countries according to whether pledges and policies are consistent with a country's fair share effort to holding warming to below 2°C. The ranking focus on pledges for 2020, 2025 and 2030. There are four categories in this rank-ing: i) role model: emission targets for this category are more ambitious than the 2°C target (worth noting that no country in the world is in this category); ii) sufficient: if all governments are in this category, then warming will be limited to 2°C; iii) medium: if all governments were in this rank, then warming is likely to exceed 2°C; and iv) inadequate: emission targets in this area are not ambitious enough and if all governments were in this category, warming will exceed 3–4°C. Table 7.2 shows CAT rating for South American countries as well as the US and China for comparison.

As Table 7.2 shows, Argentina and Chile are rated "inadequate" while Brazil and Peru are rated "medium." According to CAT, Argentina's current policies are insufficient as they will signify an emissions increase from all sectors (exclud-ing land use change and forest) by about 50% above 2010 levels by 2030. The report states that more ambitious policies need to be in place by President Mac-ri's administration to fairly reflect Argentina's capabilities.[6] Chile, also with the same rating as Argentina, has been criticised for submitting unambitious targets. The country has set two types of targets, a conditional one (meaning they will need international financial aid to meet this target), which intends to reduce GHG emissions-intensity of GDP by 35–45% compared to 2007 by 2030. The non-conditional target is a 30% reduction of GHG emissions-intensity of GDP below 2007 levels by 2030. According to CAT, Chile is on track to meet its unconditional NDC target, but will not meet its 2020 pledge and 2030 condi-tional NDC targets.[7]

In the case of Brazil, the rating of "medium" reflects the fact that the country's current policy is on track to meet its 2025 target, but will fall short of its 2030 goal. Still, CAT states that Brazil is one of the world's largest emitters, and as such its policies can have a great impact in meeting the 2°C global target. The report adds that Brazil's NDC emissions reduction targets "are at the least ambitious end of a fair contribution to global mitigation, and are not consistent with meeting the Paris Agreement's long-term temperature goal unless other countries make much deeper reductions and comparably greater effort."[8] In the case of Peru, their NDC involve a reduction by 20% below business-as-usual (BAU) by 2030 (unconditional), and a 30% reduction relative to BAU levels by 2030 (conditional on international aid). The report highlights that the main problem in Peru is deforestation, stating that emissions are projected to soar from 92.6 $MtCO_2e$ in 2010 to 159 $MtCO_2e$ in 2030 due to this activity.[9]

The debate surrounding developing countries' capability to do more with their NDCs has centred around fair share of efforts. Indeed, due to wealth and levels of productivity, some countries are more developed and have better technologies to deal with climate change, some others are more vulnerable to risk, while others have already emitted a considerable amount of CO_2 for a long period of time, which has placed them at an advantage in terms of infrastructure. This debate has influenced the UN to advocate for common but differentiated responsibilities based on respective capabilities.[10] Still, as seen by CAT's assessment, Argentina and Chile are not exhausting the capacity they have to reduce greenhouse gas emissions to the atmosphere.

Debate also surrounds the NUA. For instance Richard Friend, commenting on Views&Voices.Oxfam.org has questioned the NUA's ability to envision new models or to re-imagine the urban future. He has also highlighted the problems around leadership: "The core problem that we face in dealing with the urban challenge is a problem of governance – of politics, power and knowledge."[11] He goes on to state that in developing countries, problems of governance and corruption are hampering the delivery of much needed projects. He cites planning institutions and processes as being secretive, benefiting speculation and with no real sense of shared urban futures. Furthermore, he points out that though the NUA calls for accountability and transparency, access to information, redress or remedy are limited. However, he does see citizens, particularly in poor communities, becoming more engaged in the political process, more informed and articulate, and he believes that they will ultimately be the drivers for change that will shape the new urban agenda.[12]

The Intergovernmental Panel on Climate Change (IPCC) has also voiced similar concerns, highlighting global constraints on implementation arising mainly from limited financial and human resources, limited integration or coordination of governance, uncertainties about projected impacts, different perceptions of risks, competing values, absence of key adaptation leaders and advocates, and limited tools to monitor adaptation effectiveness. Another area of constraint includes insufficient research, monitoring, and observation and the finance to maintain them.

It is clear that the style of governance and policy implementation will be a key challenge in the future and that BAU will no longer be acceptable; fundamental change is needed. The IPCC has presented clear scenarios and provided a "global carbon budget," which limits the amount of greenhouse gases that the world can emit to 400–850 GtCO$_2$ for the period 2011–50 for a 50% chance to keep temperatures rise below 1.5°. Their studies also suggest that this budget will be consumed in ten to 20 years if the world continues at current levels. Keeping within this carbon budget must assure that efforts are shared equitable and fairly.

According to this scenario, the IPCC has conducted an extensive research of academic literature and has found that scientific evidence has high confidence in that large changes in investment patterns are needed in order to achieve substantial reduction in emissions (IPCC, 2014). This coincides with the views presented here in Chapter 5 regarding infrastructure, including Expert boxes 5.1 and 5.2. The IPCC also expressed that there is high agreement amongst researchers in that involvement of the private alongside the public sector can play a role in financing mitigation and adaptation strategies. In the case of South American governments and in terms of real estate, PPPs such as those currently being pursued by the Argentine government (see Chapter 4), can certainly provide a good pathway to reduce the burden of the state in improving living conditions for those at the bottom end of the social pyramid. However, more research is needed to measure empirically the success of these type of policies.

Academic institutions and funding agencies can play a role in influencing this change by encouraging new practices, facilitating knowledge sharing and providing evidence of successful practice. The real estate industry has the potential to lead in a new world of sustainable development if institutions encourage and empower them to do so. This could be achieved through incentives and tax breaks. Governments and delivery agents will need to shift to new modes of cooperation and forward-looking co-investment approaches that can foster the skills, innovation and technologies required in the coming decades.

7.6 How South America can help achieve the SDGs and protect real estate and infrastructure assets

According to the IPCC, climate change is projected to increase risks for people in urban areas. The organisation highlights that assets, economies and ecosystems are at risks attributed to heat stress, storms and extreme precipitation, inland and coastal flooding, landslides, air pollution, drought, water scarcity, sea level rise and storm surges. Expert box 5.2 presents the case of extreme events affecting Argentina and the investment in infrastructure of USD 158.37 million to prevent flooding in the Lujan River Basin. This signifies a large investment for the country and helps to put into context the vast amount of financial resources needed by countries in order to combat climate change.

The IPCC's models show very high confidence in the likelihood of the above mentioned events occurring, particularly sea level rise and storm surges.[13] These risks are amplified for those lacking essential infrastructure and services or living in exposed areas. As seen in Chapter 1, South America has a large proportion of slum dwellers (on average, around 30% of the total population), the vulnerability of this sector is clear, with the IPCC rating either as medium or high, the risks of floods and landslides in the region; the risk of food production and quality; and the risk of spread on vector-borne diseases (see Figure 7.1).

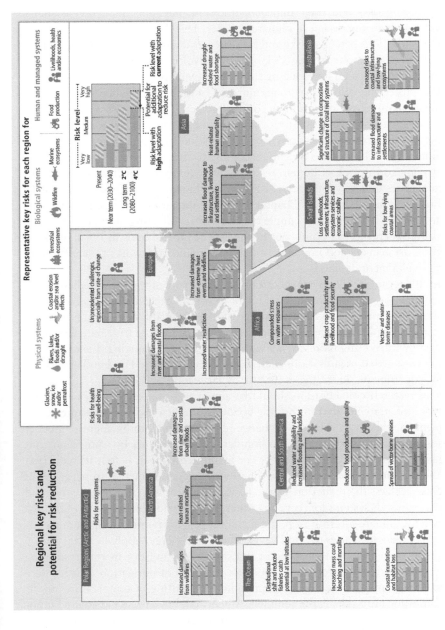

Figure 7.1 Regional key risks and potential for risk reduction

© IPCC

As seen in the graphic, current adaptation methods can reduce the risks of flooding, for example, to a low level, but this is not sufficient for the near term (2030–2040) when the risk will be back at medium level, and even worse for the long term (2080–2100) when the risk will be high.

Analysing risks and their impact factor (low, medium, high, severe) helps governments in the design of climate policies. This depends of how individuals and organisations perceive risks and how they evaluate them from an economic, social and ethical point of view. Adaptation and mitigation strategies are complementary methods used for reducing these risks over a period of time. Adaptation entails protecting vulnerable groups and assets and this can be best implemented through planning policies. Still, according to the IPCC, holistic environmental views of some communities – for example, those of indigenous groups – are not being incorporated or taken into account when designing and planning adaptation methods. The reasons for this, according to the same source, are varied but they point to the lack of human resources at government level to interact with communities and lack of research. Again the human resource shortages highlighted in Case Study 3.1 in Chapter 3 when scaling up social housing solutions from Chile to Brazil are an example of this shortage affecting most governments. Equally, Expert box 5.1 points out to the lack of professionals in Brazil with the capacity to manage and implement complex engineering works that infrastructure development demands.

In this regard, research organisations have a key role to play as mediators and as a source for human capacity and talent to fill the gap existing in government departments. International partnerships (SDG 17) can also help here, as academics across disciplines in countries such as the UK are looking to enhance research impact by interacting more closely with governments. Financing this research is the duty of global organisations such as the Green Climate Fund and at country level, government aid strategies such as Newton Fund in the UK.

This research collaboration should be aimed at a mixture of adaptation and mitigations strategies that in the case of South America and for real estate development should cover: i) land-use change reduction and improvement of greener cities to enhance carbon sink environments (for example, by using land zoning policies that protect green spaces or by imposing green belts); ii) reduction in the consumption of energy and improving energy efficiency in a variety of end-use sectors; iii) reduction in the consumption of water in urban areas and moving towards systems for recycling water; iv) reduction of emissions of health-damaging, climate-altering air pollutants by improving public transport and considerably reducing car usage; and v) decarbonising energy supply.

These strategies need cross-sector transformation and not merely restricting usage in individual sectors that can affect the need of mitigation in other sectors. For example, some countries promote the limiting of temperatures in air conditioning units during the summer (Argentina), but a more radical and innovative transformation is needed in the construction sector in terms of insulating materials and in the design sector in terms of maximising air circulation. Equally, reducing or limiting parking spaces in commercial buildings as well as implementing car entry restrictions into city centres, are certainly good temporary measures but a deeper cultural transformations is needed to drastically cut down car dependancy.

Adaptation and mitigation responses are mainly underpinned by governments and institutions, as they have and important role not only in the design of policies and management of budgets, but also in the scaling up of adaptation and mitigation strategies to communities, households and civil society. According to the IPCC (2014), researchers have found that linkages among regional, national and sub-national climate policies offer potential climate change mitigation benefits. Possible advantages include lower mitigation costs, decreased emission leakage and increased market liquidity.

In terms of specific policies, carbon pricing has been implemented in Chile in the power sector since 2014. The measure affects thermal plants with installed capacity equal or larger to 50 MW. Such plants now pay USD 5 per ton of CO_2 released. The policy aims at incentivising the sector to move towards cleaner resources. This can be seen as a good example in Chile, but research is needed to assess its impact and transferability to other countries, particularly since the IPCC states that scientific evidence is limited on carbon pricing mechanisms.

On the other hand, empirical evidence is strong in respect to taxation policies that are aimed at reducing GHG emissions, as collected tax revenues reduces the impact on the national GDP. The examples can be on fuel taxes but equally further studies could be developed in relation to land use change and land taxation for development, in order to fund much-needed infrastructure in a way that it will not affect the public purse. All the Big Five economies under study in this volume implement subsidies for house development and improvements, incorporating conditional clauses for subsidies to encourage sustainable construction can be implemented. The scientific evidence shows high confidence that economic incentives such as tax rebates or exemptions, grants, loans and credit lines, can expand the use of renewable energies and technologies as well as achieve emission reductions (IPCC, 2014).

Regulatory policies can also help to encourage sustainable development. Again there are examples of academic evidence supporting the implementation of energy efficiency standards in construction, as well as labelling programmes for buildings standards and appliances, which can help consumers to make informed decisions. In this sense, Expert box 5.4 already points to the need for encouraging and strengthening bottom-up approaches from the wider community. Sustainable livelihoods and behavioural lifestyle choices have an important impact on reducing GHG as emissions can be substantially lowered through changes in consumption patterns, adoption of energy savings measures, dietary change and reduction in food wastes.

Prior to the World Economic Forum on Latin America in Medellín, Colombia, in June 2016, Alicia Bárcena Ibarra of the Unites Nations Economic Commission for Latin America (ECLAC) outlined thoughts on how Latin America might respond to the SDGs. Her recommendations focused on the need for PPPs and establishing new legal frameworks to ensure that risk is shared evenly, with more active fiscal policies, aimed at fostering low carbon-growth paths and full employment. She cites the fact that this is the most urbanised region in the world (80%) as an opportunity to pursue innovation in public transport and traffic management, solid waste and water treatment and low-energy buildings. She also believes that big data can be used for the public good, to help decision-making, and urges collaboration between businesses, governments and civil society to learn from existing successes so as to monitor the achievements of the SDGs. However, she warned that Latin America needs to improve regional and sub-regional integration by boosting inter-regional trade and the regional value chains in environmental goods and services; this will need to be supported by strengthened financial safety nets and payment clearing systems. Finally, she draws attention to the need for greater coordination to control illicit capital flows and to apply common fiscal, social and environmental standards, to attract foreign investment and reduce predatory competition. She also suggests the creation of a digital common market that supports technology transfers, as well as a fund to purchase and license patents (essential in a knowledge economy). The real estate sector can engage in all of these recommendations.

Overall, there are strengths and weaknesses in terms of performance against the SDGs, but if the real estate sector is encouraged to move towards more sustainable patterns, then there is a good chance that the sector can influence other practice. SDG 11 (cities and human settlements) in particular does, of course, relate directly to real estate development and project promoters should therefore seize the opportunity to quickly improve development standards to meet the goals of the 2030 Sustainable Development Agenda.

Adaptation strategies are good to address current risks and can be beneficial for future emerging risks. But mitigation is the one tool that will bring the highest benefits and can substantially reduce climate change impact in the long term. South America has the opportunity to put in motion strategies and actions now, that can pave the way for sustainable development while improving infrastructure, housing conditions and reducing poverty and inequality. Integrated responses are needed, particularly looking at nexuses of water, food and energy in the context of urban planning and development, which will provide substantial opportunities for real estate investment in increasingly resilient cities.

Notes

1 UN New Urban Agenda Explainer is available at www.un.org.
2 de Paula, N. (2016) International Institute for Sustainable Development (IISD). Available at www.sdg.iisd.org/about/
3 The full report is available at www.odi.org/publications/10454-projecting-progress-sdgs-latin-america-and-caribbean
4 Access to full SDG Index and Dashboard is available at www.sdgindex.org
5 http://climateactiontracker.org
6 Full Rating Report. Available at http://climateactiontracker.org/countries/argentina.html
7 Full Rating Report. Available at http://climateactiontracker.org/countries/chile.html
8 Full Rating Report. Available at http://climateactiontracker.org/countries/brazil.html

9 Full Rating Report. Available at http://climateactiontracker.org/countries/peru.html
10 UNFCCC United Nations Framework Convention on Climate Change. Available at https://unfccc.int/resource/docs/convkp/conveng.pdf
11 Friend, R. (3rd May 2017). The New Urban Agenda in the SDGs.
12 Comment on the New Urban Agenda. Available at www.un.org/sustainabledevelopment/blog/2016/10/newurbanagenda/
13 See IPCC Report Summary. Available at http://ar5-syr.ipcc.ch/topic_summary.php

Bibliography

IPCC – Intergovernmental Panel on Climate Change. (2014) *Fifth Assessment Report*. Available at http://ipcc.ch/report/ar5/index.shtml, accessed on 10th June 2017.

UN Habitat. (2011) *Hot Cities: The Battle-Ground for Climate Change*. Avaiable at http://mirror.unhabitat.org/downloads/docs/E_Hot_Cities.pdf, accessed on 10th July 2017.

UN Sustainable Development Goals. Available at www.un.org/sustainabledevelopment/sustainable-development-goals/.

Appendix 1

 The UN Sustainable Development
Agenda's Sustainable Development Goals

Summary in terms of development.

No 1 *End poverty in all its forms everywhere.* Development will need to address the needs of the poor as well as the wealthy in society.

No 2 *End hunger, achieve food security and improved nutrition and promote sustainable agriculture.* Food security is likely to be an issue in the future as cities expand into the surrounding countryside, but there are good reasons for growing a significant proportion of a city's needs within the city itself. Development should take this into consideration and include places where food products can be grown.

No 3 *Ensure healthy lives and promote well-being for all at all ages.* Development will need to be designed with healthy lifestyles in mind, promoting active travel and leisure.

No 5 *Achieve gender equality and empower all women and girls.* The design of new developments will need to be inclusive so that all are welcome and vulnerable people feel safe.

No 6 *Ensure availability and sustainable management of water and sanitation for all.* Adequate water supply and sanitation will need to consider innovative and non-carbon-intensive systems.

No 7 *Ensure access to affordable, reliable, sustainable and modern energy for all.* Every new development creates new demand for energy, so innovation and sustainable energy systems need to be central to the planning of new communities and facilities.

No 9 *Build resilient infrastructure, promote inclusive and sustainable industrialization and foster innovation.* Infrastructure innovation in design, planning and delivery will have to be central to all future development.

No 10 *Reduce inequality within and among countries.* Much social inequality has arisen as a result of poor or non-existent planning and the rapid growth of informal settlements. It must become the responsibility for development to address this imbalance and create opportunity for the less fortunate in society.

No 11 *Make cities and human settlements inclusive, safe, resilient and sustainable.* Much crime occurs in poorly designed urban environments that make it easy for illegal activity to take place. Such places exclude much of society and good design is necessary to avoid places that nurture crime. This is a key aspect of sustainability and resilience of cities.

No 12 *Ensure sustainable consumption and production patterns.* This goal aims to promote sustainable consumption, production patterns and the management of materials that are toxic to the environment. Development should consider carefully materials that are used and their impact on people and the environment; sustainably sourced timber, for example.

No 13 *Take urgent action to combat climate change and its impacts.* If we are to live in a sustainable way, we need to evolve carbon-neutral practices. Through good design and planning, it is possible to create new development that is not carbon-intensive and which promotes non-carbon-intensive travel.

No 15 *Protect, restore and promote sustainable use of terrestrial ecosystems, sustainably manage forests, combat desertification, and halt and reverse land degradation and halt biodiversity loss.* Development can contribute to the support of ecosystems and biodiversity by providing habitats and ensuring that the systems that support the development do not damage the wider environment through their use of resources or their handling of waste.

No 16 *Promote peaceful and inclusive societies for sustainable development, provide access to justice for all and build effective, accountable and inclusive institutions at all levels.* The very design of new development can influence whether a place is inclusive with social institutions that support and foster peace and better lifestyles for all; new development must consider how it is able to contribute to this objective (though this objective is abstract, it is not intangible).

No 17 *Strengthen the means of Implementation and Revitalize the Global Partnership for Sustainable Development.* It is clear that single client or delivery projects are more and more an outmoded model and that partnership working is more inclusive as well and economically expedient.

Appendix 2

INDCs data, World Resources Institute's CAIT Climate Data Explorer, http://cait.wri.org/pledges/

Country	Greenhouse gas emissions including land use change and forestry (LUCF) 2012 (MtCO$_2$ equivalent)	INDC summary	Mitigation contribution type	GHG type
Argentina	405.03	"Argentina's goal is to reduce GHG emissions by 15% in 2030 with respect to projected Business-As-Usual (BAU) emissions for that year. The goal includes, inter alia, actions linked to: the promotion of sustainable forest management, energy efficiency, biofuels, nuclear power, renewable energy, and transport modal shift. The criteria for selecting the actions include the potential for reducing/capturing GHG emissions and associate co-benefits, as well as the possibility of applying nationally developed technologies." "Argentina could increase its reduction goal under the following conditions: a) Adequate and predictable international financing; b) support for transfer, innovation and technology development; c) support for capacity building. In this case, a reduction of 30% GHG emissions could be achieved by 2030 compared to projected BAU emissions in the same year."	GHG target	Baseline scenario target
Bolivia	136.47	Bolivia presents its contribution "in two dimensions: one linked to the structural solutions, and other results and national actions within the framework of holistic development."	Non-GHG target and actions	Not Applicable
Brazil	1823.15	"Brazil intends to commit to reduce greenhouse gas emissions by 37% below 2005 levels in 2025."	GHG target	Base year target

Chile	93.74	"Carbon intensity target, not including the LULUCF sector: a) Chile is committed to reduce its CO_2 emissions per GDP unit by 30% below their 2007 levels by 2030, considering a future economic growth which allows to implement adequate measures to reach this commitment[1]. b) In addition, and subject to the grant of international monetary funds[2], the country is committed to reduce its CO_2 emission per GDP unit by 2030 until it reaches a 35% to 45% reduction with respect to the 2007 levels, considering, in turn, a future economic growth which allows to implement adequate measures to achieve this commitment. Specific contributions to the LULUCF sector: a) Chile has committed to the sustainable development and recovery of 100,000 hectares of forest land, mainly native, which will account for greenhouse gas sequestrations and reductions of an annual equivalent of around 600,000 of CO_2 as of 2030. This commitment is subject to the approval of the Native Forest Recovery and Forestry Promotion Law. b) Chile has agreed to reforest 100,000 hectares, mostly with native species, which shall represent sequestrations of about 900,000 and 1,200,000 annual equivalent tons of CO_2 as of 2030. This commitment is conditioned to the extension of Decree Law 701 and the approval of a new Forestry Promotion Law." [1] This commitment assumes a growth rate for the economy similar to the growth path the country has experienced in the last decade, except for the most critical years of the international financial crisis (2008–2009). [2] This commitment assumes a growth rate for the economy similar to the growth path the country has experienced in the last decade, except for the most critical years of the international financial crisis (2008–2009). In addition, for the purposes of this commitment, an international monetary grant shall be deemed any grants which allow to implement actions having direct effects on greenhouse gas emissions within adequate time frames.	Intensity target GHG target and non-GHG target
Colombia	199.68	The INDC also includes an adaptation contribution. The Republic of Colombia commits to reduce its greenhouse gas emissions by 20% with respect to the projected Business-as-Usual Scenario (BAU) by 2030. "Subject to the provision of international support, Colombia could increase its ambition from 20% reduction with respect to BAU to 30% with respect to BAU by 2030." The INDC also includes a section on adaptation.	Baseline scenario target GHG target

(Continued)

(Continued)

Country	Greenhouse gas emissions including land use change and forestry (LUCF) 2012 (MtCO$_2$ equivalent)	INDC summary	Mitigation contribution type	GHG type
Ecuador	138.16	"Ecuador intends to reduce its emissions in the energy sector in 20.4–25% below the BAU scenario. However, a potential for reducing emissions even further in the energy sector, to a level between 37.5 and 45.8% with respect to the BAU baseline has also been calculated. This potential could be harnessed in light of the appropriate circumstances in terms of availability of resources and support offered by the international community. This is a second scenario dependent upon international support and will translate into a per capita emissions reduction in 2025 of 40% below the BAU levels."	GHG target and non-GHG target	Baseline scenario target
Paraguay	110.98	Please note that the INDC was submitted only in Spanish. WRI did its best to translate the INDC language. If any errors are identified, please contact us at wcait@wri.org "20% de reducciones en base al comportamiento de las emisiones proyectadas al 2030. Meta Unilateral: 10% de reducción de emisiones proyectadas al 2030 Meta Condicionada: 10% de reducción de emisiones proyectadas al 2030" 20% reduction relative to projected emissions by 2030 Unilateral Target: 10% reduction from projected emissions by 2030 Conditional Target: 10% reduction from projected emissions by 2030	GHG target	Baseline scenario target
Peru	159.5	"The Peruvian iNDC envisages a reduction of emissions equivalent to 30% in relation to the Greenhouse Gas (GHG) emissions of the projected Business as Usual scenario (BaU) in 2030. The Peruvian State considers that a 20% reduction will be implemented through domestic investment and expenses, from public and private resources (non-conditional proposal), and the remaining 10% is subject to the availability of international financing[1] and the existence of favorable conditions (conditional proposal)." [1]It should be noted that Peru will not assume conditional commitments that might result in public debt. The INDC also includes a section on Adaptation.	GHG target	Baseline scenario target

Uruguay	14.76	GHG target	Uruguay's INDC covers both mitigation and adaptation components. The mitigation component of Uruguay's INDC is "sorted by gases" and covers nine quantitative contributions that can be met "with domestic resources" and "with additional means of implementation," all to be achieved by 2030. The adaptation component of Uruguay's INDC outlines ten adaptation actions that Uruguay expects to accomplish by 2030, "with the support of external means of implementation." See the INDC for more information.	Uruguay has outlined eight intensity targets and one fixed level target.
Venezuela	396.74	GHG target	Please note that the INDC was submitted only in Spanish. WRI did its best to translate the INDC language. If any errors are identified, please contact us at *wcait@wri.org* "Sin embargo, como parte de las políticas establecidas en el plan de desarrollo económico y social del país, Venezuela se propone implementar un Plan Nacional de Mitigación en conjunto con un Plan Nacional de Adaptación. El Plan Nacional de Mitigación apuntará a la reducción de las emisiones del país en al menos un 20% para 2030 en relación al escenario inercial, entendido este como un escenario hipotético en el cual no se implementa el plan. El grado en que se alcance esta meta dependerá del cumplimiento de los compromisos de los países desarrollados en cuanto a provisión de financiamiento, transferencia de tecnología y formación de capacidades de acuerdo al Artículo 4.7 de la Convención." However, as part of the policies set forth in the plan for economic and social development of the country, Venezuela intends to implement a National Mitigation Plan in conjunction with a National Adaptation Plan. The National Mitigation Plan will aim to reduce the country's emissions by at least 20% by 2030 in relation to the baseline scenario, understood as a hypothetical scenario in which the plan is not implemented. The degree to which this goal is achieved will depend on the fulfillment of the commitments of developed countries in terms of provision of finance, technology transfer and capacity building pursuant to Article 4.7 of the Convention.	Baseline scenario target

(Continued)

(Continued)

Country	Greenhouse gas emissions including land use change and forestry (LUCF) 2012 (MtCO$_2$ equivalent)	INDC summary	Mitigation contribution type	GHG type
China	10684.29	China has nationally determined its actions by 2030 as follows: • To achieve the peaking of carbon dioxide emissions around 2030 and making best efforts to peak early; • To lower carbon dioxide emissions per unit of GDP by 60% to 65% from the 2005 level; • To increase the share of non-fossil fuels in primary energy consumption to around 20%; and • To increase the forest stock volume by around 4.5 billion cubic metres on the 2005 level.Moreover, China will continue to proactively adapt to climate change by enhancing mechanisms and capacities to effectively defend against climate change risks in key areas such as agriculture, forestry and water resources, as well as in cities, coastal and ecologically vulnerable areas and to progressively strengthen early warning and emergency response systems and disaster prevention and reduction mechanisms.	GHG target and non-GHG target	Intensity target, Trajectory target
US	5822.87	The United States intends to achieve an economy-wide target of reducing its greenhouse gas emissions by 26%-28% below its 2005 level in 2025 and to make best efforts to reduce its emissions by 28%.	GHG target	Base year target

Appendix 3
New Urban Agenda

Key agreements

1 **Provide basic services for all citizens**
   These services include: access to housing, safe drinking water and sanitation, nutritious food, healthcare and family planning, education, culture and access to communication technologies.

2 **Ensure that all citizens have access to equal opportunities and face no discrimination**
   Everyone has the right to benefit from what their cities offer. The New Urban Agenda calls on city authorities to take into account the needs of women, youth and children, people with disabilities, marginalised groups, older persons, indigenous people, among other groups.

3 **Promote measures that support cleaner cities**
   Tackling air pollution in cities is good both for people's health and for the planet. In the Agenda, leaders have committed to increase their use of renewable energy, provide better and greener public transport, and sustainably manage their natural resources.

4 **Strengthen resilience in cities to reduce the risk and the impact of disasters**
   Many cities have felt the impact of natural disasters and leaders have now committed to implement mitigation and adaptation measures to minimise these impacts. Some of these measures include: better urban planning, quality infrastructure and improving local responses.

5 **Take action to address climate change by reducing their greenhouse gas emissions**
   Leaders have committed to involve not just the local government but all actors of society to take climate action, taking into account the Paris Agreement on climate change which seeks to limit the increase in global temperature to well below 2 degrees Celsius. Sustainable cities that reduce emissions from energy and build resilience can play a lead role.

6   **Fully respect the rights of refugees, migrants and internally displaced persons regardless of their migration status**
Leaders have recognised that migration poses challenges but it also brings significant contributions to urban life. Because of this, they have committed to establish measures that help migrants, refugees and IDPs make positive contributions to societies.

7   **Improve connectivity and support innovative and green initiatives**
This includes establishing partnerships with businesses and civil society to find sustainable solutions to urban challenges.

8   **Promote safe, accessible and green public spaces**
Human interaction should be facilitated by urban planning, which is why the Agenda calls for an increase in public spaces such as sidewalks, cycling lanes, gardens, squares and parks. Sustainable urban design plays a key role in ensuring the liveability and prosperity of a city.

Index

Page numbers in italic indicate a figure and page numbers in bold indicate a table or graphic.